Improve Your TECHNICAL COMMUNICATION

Improve Your TECHNICAL COMMUNICATION

John S. Campbell

Former member of the faculties of:

California Institute of Technology
California State Polytechnic College
Claremont College
Long Beach State College
Grantham School of Engineering

GSE PUBLICATIONS

2000 Stoner Avenue, Los Angeles, California 90025

Library of Congress Cataloging in Publication Data

Campbell, John Scott.
    Improve your technical communication.

    Consists of selected lessons in technical communication, first written as correspondence lessons, making up a small part of a correspondence course in electronics engineering.

    1. Technical writing. I. Title. II. Title: Technical Communication.
T11.C28   808'.066'6021   76-8493
ISBN 0-915668-26-2

FOREWORD

The "chapters" of this book are called *lessons,* because they were first written as correspondence lessons, making up a very small part of a correspondence course in Electronics Engineering. In that course, the lessons in *technical communication* are not presented consecutively but, instead, are interspersed among other lessons in electronics, mathematics, physics, and computer science.

I believe there are many students, technicians, engineers, and scientists who have no need for the entire correspondence course that *do,* nevertheless, need to improve their skills in technical communication. Many such persons must write effective reports as a regular part of their work. Many others, who have technical knowledge and experience, realize that the ability to write technical manuals, articles, proposals, etc., can result in expanded occupational opportunity.

Therefore, these selected lessons in Technical Communication are being published here, not as correspondence lessons but in book form, for all who can benefit from them. The sequence of lessons included is independent of the correspondence course, and the reader should ignore references to "the course" and "the Series."

Donald J. Grantham
President
GSE Publications

CONTENTS

LESSON TC-1
Introduction to Technical Communication

Our lives are certainly different from those of our grandparents, in that the earth constantly grows "smaller" and outer space "nearer." This is largely a result of developments in communications electronics and of other great strides made in engineering. Scientific discoveries are followed by the invention of ingenious devices which are *engineered* into systems that result in our technological civilization becoming more and more technological. Without engineers and technicians, these systems could not be developed. And, come to think of it, without the *communication of ideas through words,* the whole big project would never have gotten off the ground. That is, discoveries and inventions cannot be made and developed without much talking and/or writing being done—technical communication being accomplished —in the process.

To write and talk, in the process of technical development and otherwise, we must use words carefully chosen and arranged so that precise communication occurs. Technology cannot survive on a diet of fuzzy, inexact attempts at communication. Engineers and technicians must communicate among themselves with precision and accuracy if technological efforts are to succeed, and they must communicate with people outside their fields if they are to succeed as social human beings.

As a technician, technologist, or engineer, you should be able to skillfully use your native language in the communication of your own thoughts to others, and in grasping thoughts written and spoken by others. The lessons provided in this book are designed to improve your ability to communicate in general, and to communicate with other engineers and industry members in particular.

Standards of Good Writing

The basic units of language are words. We may subdivide words to get smaller sub-units, and we may string words together to make meaningful phrases, clauses, sentences, paragraphs, and compositions. So in studying the units of language, where do we begin? Actually, you began your study when you were a baby, and this lesson (being written for many people of many different backgrounds) cannot begin at the right place for everybody. We *could* estimate where our *average* student should begin, and then "dive in" but, instead, we have begun at a point that is not too advanced for *any* student even though this lesson may be just a review for most.

Most of our students, unfortunately, have not mastered the tools of the writing craft for either non-technical or technical communication. But all of them do have a lot of knowledge which applies toward this mastery. What do we mean by "tools of the writing craft"? Well, you have guessed that we do not mean pen and paper. And we must admit that this phrase is not the kind of precise language we have spoken of in the introductory paragraphs of this lesson. By tools of the writing craft, we could mean *words.* Or we could mean grammar, rhetoric, and composition, or we might think of these latter "items" as the three tool boxes in which the real tools are found:

Grammar deals with the relationship between different kinds of words used together in making sentences. That is, grammar deals with *parts of speech,* and the rules regulating which forms of words are used under which conditions.

Rhetoric deals with diction and style—that is, with the choice of words and their arrangement in the writing of good prose (ordinary language rather than poetry).

Composition deals with putting together the parts to form the whole. Thus in writing, composition is the forming of sentences from words, and the forming of paragraphs from sentences, and the forming of a "piece of writing" (composition) from the paragraphs. Compositions are referred to as *essays, narratives, articles, lessons,* etc. The *whole* piece of writing is a composition.

The tools of grammar may be thought of as the *rules of grammar,* the tools of rhetoric as the *rules of rhetoric,* and the tools of composition as the *rules of composition.* All of these taken together make up the *rules of good English.* Such rules are not purely arbitrary but, instead, are outgrowths of generations of "good usage." "Good usage" means the usage that is common among accepted leaders in the art of language communication—the leading authors in various fields—journalists, novelists, etc.

Your Dictionary

In this Series we normally do not suggest that you buy a particular kind of book. In general, the books that you need are supplied by the School. But since most students already own a good dictionary we do not supply this book. A good dictionary is a must in the home of every "educated" person. If you do not already have regular access to one, be sure to acquire one right away. Get the best dictionary you can afford. A poor one will hinder rather than help you, because in looking for things it does not contain you will waste your time.

A dictionary that we would call "good" should have a preface (or introduction) and should have other special sections such as standard abbreviations, geographical names, biographical names, a general guide to pronunciation, etc. Of course the main part of a dictionary (as far as the number of pages in concerned) is the vocabulary—the alphabetical list of words with their pronunciations and definitions. The definitions should be concise but complete and should be easy for you to understand. The system of phonetic symbols used in the dictionary you select should be one that you are already familiar with, or the dictionary should carefully explain (and you should study) the system used.

Then, once you have your dictionary, use it! It can be a great help to you in many ways—not only in studying this Series but also in becoming a better-educated (and higher-paid) person. When you hear or read a word you don't know or a word you are not sure of, make a mental (or written) note and at your next opportunity look it up. When you look it up, study and reflect over the word for a few moments. Don't stop after looking at the definition. Look at the phonetic symbols which tell you its proper pronunciation. Note the syllable or syllables upon which the spoken accent(s) fall(s). Say the word aloud a few times, and listen to the rhythm and the ring of it.

By keeping a little list of words, both technical and non-technical, that you are working on (dropping each word as it is mastered and adding each new word as it is discovered), you will increase your vocabulary and perhaps have some pleasure in the process. For most of us it is pleasant to know we are learning, and increasing your mastery of words can be very profitable in the long run.

Dictionaries are written by people called *lexicographers,* also sometimes called *dictionarists.* Of course one lexicographer does not write a whole dictionary. Usually, one is assigned to some portion of the vocabulary such as, for example, all words that begin with a. In this manner a group of lexicographers produce the whole vocabulary. (Lexicographer is pronounced, lex-uh-KAHG-graf-uhr.)

The vocabulary information placed in the dictionary is obtained through research. That is, each lexicographer does a great deal of reading of all kinds of works. In doing this reading, he compiles a list of sentences in which each vocabulary word is used. After collecting a great number of sentences from many authors for a given vocabulary word, he draws conclusions from those sentences as to the meaning or meanings of the word. As you can see, he derives the meaning from many different writers, writing about different subjects at different times and places. It is the lexicographer's task to condense what he discovers about the word into a concise dictionary definition that will satisfy almost everybody who wishes to use the word in any of many different circumstances. Perhaps you have noticed that most words have several different definitions given in a good dictionary. This is necessary because in most cases there are different ways that a word may be properly used. Usually the context in which you find a word indicates which of several meanings is intended. The lexicographer has listed the different meanings in the dictionary because he found reasonably common usage of all of these different meanings.

A dictionary does not say that a particular definition or pronunciation is correct; it merely reports what its lexicographers found being used. Usage makes the language. Lexicographers merely report usage. We often hear it said that such and such a pronunciation or definition is correct because the dictionary says so. It is important to realize that, strictly speaking, such a comment does not tell the truth. However, if we are using the word "correct" to mean "commonly used," then this kind of comment can be true.

English is a constantly changing language. Old words are being dropped from use and new words are coming into use all the time. The language now is not the same as it was last year at this time. Also, it is not the same as it will be next year at this time. So the fact that a certain word is not in the dictionary does not prove that it does not exist. It is impossible for the most diligent lexicographers to keep dictionaries absolutely current. No one knows how many words there are in the English language at any given time. However, scholars estimate that there are now more than one-half million English-language words.

As an educated person and especially as a technical writer, you must learn to use words in accordance with the accepted norm, as recorded in good dictionaries. Since your object as a technical writer is to *inform* the reader, you must be careful to use words with the intention to have them mean what he will likely understand them to mean. That is, in choosing and using words you must operate within a *common frame of reference* with your intended reader. Thus, in using a dictionary to find definitions, you must apply enough knowledge of the subject to prevent your selection of the wrong definition of a given word. For example, both dentists and mathematicians use *matrices*, but the connotation of the word is quite different between the two different professionals. Only a very complete dictionary will give you both of these meanings, in this particular case.

You should never jump to a firm conclusion of what is meant by a given word in a given sentence unless or until you are absolutely sure you have the same frame of reference as the writer of that sentence. The realization of what frame of reference is being used is essential at both ends of a communication—for the writer and reader alike. Keep in mind that frame-of-reference care must be taken *in addition* to the use of a dictionary, *not instead* of its use.

TEST TC-1

1. Why should an engineer be able to write, and practice the writing of, "good English"?

2. What is meant by "correct usage" as it applies to the English language?

3. What is the difference between <u>grammar</u> and <u>rhetoric</u>, and why are both important in the writing of compositions?

4. Discuss in 100 words or less what is meant by "rules of good English".

5. Discuss what is meant by "frame of reference" and give an example of why this concept is important to a technical writer.

6. Explain in your own words (75 words or less) how dictionaries are written.

7. Look up the word "semantics" in your dictionary, and after studying the definitions given write a brief discussion as to whether a writer should have a keen sensitivity in the field of semantics, and why.

8. List the names of the different <u>sections</u> found in your own dictionary.

9. Which <u>section</u> of your dictionary deals with pronounciation? What are phonetics, and why is it necessary to thoroughly understand the particular system of phonetics used in your own dictionary?

10. In looking through and studying the non-vocabulary sections of your dictionary, have you discovered anything new about what dictionaries have to offer? Which of the non-vocabulary sections do you find most interesting? Why?

LESSON TC-2
The English Language and Communication

Introduction

History may be divided into periods characterized by the emphasis given to particular activities. Intellectual inquiry characterized the classic age of Greece, while the times of the Roman empire were dominated by the organization of human effort, typified by the bureaucracy which was invented and developed by Augustus and his successors. The renaissance (literally, rebirth) was a time of re-evaluation of philosophic thought in the light of experimental inquiry. An inevitable consequence of the scientific discoveries of the 1600's and 1700's was their application to practical uses, which gave rise to the *Industrial Revolution*.

The industrial revolution continues today, but its original emphasis on the replacement of human power by mechanical power has given way to a broader goal — to the release of human beings from the necessity of performing physical and mental tasks that can be done by artificial devices. Many names have been offered to describe this second stage of the industrial revolution. It has been said that we are living in the age of computers, and that we are in the midst of an *information explosion*. More and more, systems and processes which formerly operated independently are being brought into coordination with each other, and knowledge gained by one person is more quickly made available to everyone. All of these activities require the quick and efficient exchange of information, and so it is becoming increasingly clear that our age is, among other things, an *age of communication*.

The engineer is particularly concerned with communication. In his work he must have access to current practice, and, since his only product is information he develops on his designs, he must employ techniques of communication to deliver the fruits of his labors to his client. The kind of information generated and used by engineers is more complex than that used in the exchange of ideas in casual contacts between people. Complicated concepts are used; many of them are difficult to visualize; much of the information is quantitative in nature. Distinctions are often needed between different ideas that are almost the same. Sometimes sharp legalistic definitions are needed, some of which are very different from the meanings implied and inferred for the same terms when used in everyday language.

This is the second of a series of lessons designed to acquaint you with techniques for communicating ideas, with particular emphasis on technical information. We might have called the series "Technical English" or "Technical Writing", but we chose the general title of "Technical Communication" for several reasons. Chief among these is that this name emphasizes the overall purpose of the techniques we shall discuss. It also warns us that the English language is not the only method used to communicate. Drawings, graphs, diagrams of many kinds and mathematical expressions also have their place in building a concept in the mind of the recipient, and we shall include these in our discussions where appropriate. In general, however, our main emphasis will be on *the use of words to convey ideas*.

This introductory lesson discusses the use of language in human society in general, and the particular demands made on language by the needs of the engineer. It discusses the various types of technical writing that are used in industry, and shows how style and vocabulary are varied to suit these types, and to have maximum effect on the reader whose technical background and vocabulary may be very different from that of the writer. Succeeding lessons will expand on the ideas introduced in this lesson, with examples and opportunities to put theory into practice.

What is Language?

The English language is one among several thousand distinct languages now in use by mankind. We are fortunate in having it as our native tongue, because it is one of the *great* languages, capable of infinite variety of expression and capable of being the vehicle for a rich literature. It is also the most widely used language on earth and it would seem to us that, barring possible results of wars and/or natural disasters, it is destined to be a principal component of a future world language.

All natural languages originated in remote prehistory, probably several hundred thousands of years ago. We can only guess as to the nature of their origins. Cries of warning, sounds indicative of affection and calls to food, which animals use today, are thought by some anthropologists to be the prototypes for the sonnets of Shakespeare. In any event, by the time civilization began some ten thousand years ago, there existed many richly-endowed languages.

At first languages used only sound, but eventually the major cultures developed written symbology. Today only primitive tribal languages lack writing.

The original purpose of language was purely communication. Users, however, soon found that when they formulated thought into words, they used the words in the process of thinking. That is, thought which originally was purely a series of sensory images, came to be expressed also in words. Our facility in rapid verbal expression comes about because the thought back of the expression is also composed of words. In other words, we have only to read off what is already registered in the mind. The habit of verbalizing thought is so strong that many, particularly old people, get into the habit of talking to themselves. Such a practice is not a sign of senility, but merely indicates how fully symbols can take the place of reality in the human mind.

As languages developed to match the increasing complexity of human cultures, they took on the ability to express much more than factual information. Different words were used by a parent addressing his child, than by a person addressing another of his own age, or speaking to a social superior. In European languages there is a *familiar pronoun* for "you" and also a *polite form* — *du* and *Sie* in German, *tu* and *vous* in French. In English the familiar form, *thou*, has become obsolete. In Japanese the whole style of speech is altered, depending on the relative stations in life of speaker and listener. This complication, which can never be fully mastered by a foreigner, makes possible great subtlety and richness in literary expression, and gives to conversation among cultured Japanese a very great elegance.

English also is capable of subtlety in style, and the expert user of this language can and should take account of the *image* he desires to create of himself in the mind of the intended listener or reader. The careless speaker or writer who uses bad grammar, crude or slang expressions, or a limited vocabulary full of cliches, creates in the mind of an educated audience the image of a stupid boor. On the other hand, if the intended audience is composed of people of limited education, correct grammar and elegant style and use of large vocabulary will create the image of a snob. The user of English should have at his command the vocabularies and styles appropriate to any type of listener or reader, and be able to use these deliberately and intelligently.

Consider an example. A casual comment heard frequently is, "It looks like rain." Expressed in this neutral fashion, the remark gives a bland impression of an average citizen; it is the type of speech that receives little attention and is usually intended as a fill-in. This remark could be expanded to, "There's a cold front expected by 5 PM, with up to 2 inches of rain." This statement contains definite information that can be acted upon — get inside by 5, prepare for a heavy downpour. For the technically minded, the words "cold front" tell things about the nature of the storm — dropping temperature, winds, abrupt start of the rainfall and gradual tapering off. Besides presenting information that arrests attention, this comment tells something about the speaker. It tells that he is concerned with accuracy and detail, that he has an organized mind and that he is worth listening to. Even the dullard is apt to have his attention caught by the figures quoted, and the perceptive listener will be attracted to the speaker as a person of intelligence.

Another way that it could be said is, "There are strong indications of precipitation in the foreseeable future." This statement says little in many words, but tells something about the speaker. He is either a shallow-minded person in love with words that he misuses, or the whole thing is a put-on. The words and style bring to mind a fussy old gentleman with an umbrella.

Since this lesson is addressed to readers interested in technical communication, the second of the three examples is preferred. Were this a text on English composition in general, perhaps the first example would be preferred because the scientific side of weather is not especially interesting to the average man.

With this initial probe into the nature of language, let us now examine in a little more detail the way it is used.

Words and Phrases

The English language is built up out of words. A word is a unit of meaning, written with spaces about it and generally spoken with a perceptible pause before and after. There are several different *types* of words, as we shall learn in the next lesson. Some words stand for definite things: cat, house, sunset, love, sickness — the *nouns*. Other words describe attributes of things: red, large, pretty, dirty — the *adjectives*. Action in speech is provided by the *verbs* — go, run, sing, argue. Verbs in turn are qualified by *adverbs*, which perform the same service for verbs that adjectives do for nouns — he runs *fast*, he went *there*, he was *hardly* happy with it. There is a group of words that act to join words together: and, but, or yet — the *conjunctions* — and words like under, before, in — the *prepositions*. Speech and writing are done by combining these and other kinds of words into groups that create the desired meaning. Such a group is called a *sentence*. By custom, each sentence conveys the meaning of an action carried out, and each sentence must identify the agency acting, or *subject*, the action by a *verb*, and the thing acted upon, the *object*. For example, "The dog chased the cat" identifies the *dog* as subject, *chased* as the action verb, and the *cat* as the unwilling object. Sometimes the object can be omitted — "The dog barked" — when the cause of the action is unknown or not of interest. If either the subject or verb are left out, however, we do not have a sentence, but have only a fragmentary phrase.

Young children build their sentences out of individual words. As experience in the use of language grows, the speaker or writer tends to select groups of words as units. Such practice saves mental effort, but it usually results in less effective expression, since the word-groups or expressions tend to be over-used until they become cliches. Most cliches were at one time considered to be forceful or humorous, but with constant use and changing customs, their original pungency fades out and all literal meaning may be lost. "Raining cats and dogs," "cold as ice," "be a wet blanket," "all work and no play," "quick as a wink," "cute as a bug's ear" — are all expressions which convey no information except that the speaker or writer is lazy, and the result is that listeners or readers tend to discount what he says, figuratively turning off their hearing aids, or skimming rapidly through the written material. When the reader is pressed for time, as readers of technical reports frequently are, the use of space-consuming cliches causes annoyance and fatigue, and seriously reduces the effect of the report.

Cliches are frequently used in speech to gain time in thinking what to say next. They are perhaps a shade better than the "ah's" indulged in by some speakers, but their main effect on a careful listener is to create an impression of diluted thought. "In these troubled times," "I view with alarm," "for a considerable period of time" — are favorites with second-rate speechmakers who fill in meaningless words as padding. One of the worst offenders is "consensus of opinion" which is redundant, since *consensus* itself includes the idea of several opinions combined.

The only justified use of cliches is in fictional dialog, where the phrases serve to establish a character as a person of little originality and feeble intellect.

Very old people tend to speak from habit rather than from directed intent, and their speech contains both common cliches, and expressions of their own invention that they like. To an aged southerner, there may be only "damnyankees," while others (perhaps with some reason) always combine politician with crooked.

Some of the best examples of sparkle in English are made by giving an old expression a new twist. Examples: pornography = pay dirt; birth control = multiplication tabled; a kibitzer described as a person with an "interferiority" complex. The worn-out sporting expression, "sock it to me," was the basis for four variants: the electrician whose slogan was "socket to me"; the polio clinic, "Salk it to me"; the pawnshop, "hock it to me"; and the Japanese stripper named "Sockatumi Baby." Old cliches of negation take on life when combined with sports in which they have a double meaning. For example, one who doesn't like tennis says, "It's not my racket"; the non-swimmer says, "No tanks"; and the man bored by wrestling says, "It gets me down."

Cliches are usually examples of figurative speech — that is, expressions which seek to illuminate a thought by comparing with something else. Each cliche was probably at one time very witty, but then lost its punch through overuse. Ingenious speakers and writers are constantly inventing new phrases which are just as clever as any cliche in its youth, and which have the advantage of being related to our own times and customs.

Two cliches can be combined in a related manner to bring both to life again. For example, "The quickest way to lose your shirt is to put too much on the cuff."

When two meanings are attached to the same or similar words, there are infinite possibilities for new combinations, such as is the case when a butcher advertises "never a bum steer."

Definitions that twist meanings are also appreciated. Examples: Wench = something that will turn the head of a dolt. Alimony = billing without cooing. Coffee = break fluid. Hyde Park (the Roosevelt home) used as a name for a nudist camp.

Local technical terms can be turned to good use, as in the seafood restaurant near an atomic energy plant which advertised "Fission Chips." Your writer is waiting to see a restaurant called the "Tower of Pizza."

The alert speaker or writer can often find new meanings and combinations of words like the above which will add spice and interest to speeches, and even to reports, if used with care.

Variants of English

We are all familiar with the fact that people in different parts of the country speak differently, and that people in different social and economic levels also have their own vocabularies and styles of speech. Even greater variations exist between some of the different nations whose native tongue is English — England, the United States, and Australia.

Speech variations are of two kinds: variations in the pronunciation of words, and differences in meanings. In America we are familiar with regional accents — the nasal New England twang, the Texas drawl, or the famous Southern Accent which had its origin three centuries ago when Elizabethan English, transported to the isolated plantations of the South, escaped the changes that occurred in the north and in the mother country. The Southern Accent has been exported to other parts of the country by Negroes, with the result that many uninformed people call it a "Negro accent" and even attribute its musical intonation to differences in mouth and throat construction. This ridiculous idea is quickly dispelled on hearing the impeccable Oxford English of the Jamaican Negro, or the French spoken in Haiti.

Since our main concern is with written English, we cannot give more time to the fascinating study of variations in our spoken language. We can remark however, that if you are speaking to a group which has particular habits of speech, you may make them feel more at ease if you can (without obvious affectation) slant your own spoken style towards theirs. This should not be overdone, and if you feel insecure yourself, it is better not done at all, since a phony accent may be resented.

On a regional basis, many words derive from other languages spoken by early settlers. In the Southwest we use the word "playa" (beach in Spanish) for the dry lakes in the desert, and "arroyo" for a small stream, as well as "enchilada" for the Mexican food. We speak of American cowboys, but call the Argentine equivalent "vaquero" to show the national distinction. In the New York area first settled by the Dutch, many place names still reflect the Dutch language, such as "kill" for stream. (There was a town whose original name was Horskill, until the inhabitants found out what that meant in Dutch!) The word for a small stream is typical of the variety which one encounters in the United States. In different parts of the country it is called a creek, run, draw, branch, or brook, to mention only a few variants.

Slang

The major schism in the English language is that which exists between "standard" English (as used in cultured society, in formal writing, and in conversation between educated people) and the colloquial English used in daily conversation among the majority of people. Colloquial English is characterized by a vocabulary limited to the words essential in discussing everyday situations. A word like "schism" would never be used in ordinary conversation; "break" or "division" would be more likely; but a longer expression, such as "People speak two different kinds of English ..." would be most likely.

A major characteristic of colloquial English is found in its use of a whole vocabulary of words not recognized as Standard English at all — the expressions called *slang*. We all recognize what constitutes slang. Part of it is the result of incorrect grammar — "aint," "He done good," or double negatives like "I ain't done nothing." Much slang consists of new words, usually composed from older expressions, or old words with new meanings assigned. "Neat" means anything good, "split" means to depart (split away from the rest), "like" is employed as a general prefix to any remark in the current hippie lingo. Two enduring slang words are contained in the following story which illustrates slang's tenacity:

Teacher (at start of English class session): Now children, there are two words I never want to hear in this class. One of them is *swell*, and the other is *lousy*.

Kid: OK, 'teach,' what are the words?

Some slang endures, while other such expressions eventually fade away. In the 1950's, "zoot" was *the* word for good; today teenagers have never heard of it. In the early 1900's, it was witty to exclaim, "Twenty-three skiddoo!"; now we have no idea what it meant nor the significance of the number 23.

If we examine the matter of slang from a historical viewpoint, we will make an interesting discovery. Slang is actually the front-line of the change that constantly goes on in English, and which is necessary to keep the language alive and responsive to changing culture. New words are not invented by scholars, but arise among the people who have need for them. Every year hundreds of expressions achieve popularity mainly because of a humorous connotation. Most of these are soon forgotten, but a few achieve a permanent place in the language because the need for them continues. These newcomers are classed as slang for periods from a few years to as long as several centuries, and then they slip quietly into the vocabulary of standard English. A casual browsing through a good dictionary which gives origins or words and examples of their obsolete meanings in past times, will show that very many of our most respected words were once regarded as funny or even bawdy.

For example, consider the following:

snafu. This word, meaning confused or disorganized, was concocted during World War II as a parody on military abbreviation, standing for the phrase "Situation Normal, All Fouled Up." (In the original, the fourth word was different.)

guy. This word for a man is no longer considered strictly slang, but informal standard speech. It arose out of the famous gunpowder plot to blow up the Parliament Building in London, for which one Guy Fawkes was hanged. For a time after this, the execution date was commemorated as Guy Fawkes Day, with an effigy of

Guy being burned. Men who bore a fancied resemblance to Guy's effigy were humorously called "guys."

gibberish. Today this word refers to meaningless talk, and is entirely standard and respectable. It came about as follows. Around 500 years ago, wandering bands of people of Hindu origin, speaking a language they called Romany, appeared in Europe. From their dark skins, common people assumed that they might be Egyptian, and from this came the word "gypsy," and the word "gyp" to describe their trading practices. From the same source, their language was referred to as "egyptish" which later became "gibberish."

rubber. This material was discovered by Columbus on his second voyage to the new world. He called it "caoutchouc" from the Indian pronunciation, and this name is used today in many languages, such as French. The only use found for the material was as an eraser for pencil marks.In 1780 the great British chemist, Joseph Priestly, tried without success to find other uses, and finally in disgust said that it was only good as a rubber to rub out mistakes. This nickname was thought very funny, and it stuck in the English language.

gossip. This standard word was slang in the days before the Norman Conquest when English was still composed of Celtic and Saxon words without the French additions that came after 1066. In Saxon, a relative was a "sibb" (from which the word, *sibling*, meaning brother or sister, is derived). Baptism of an infant required the appointment of godparents, or "godsibbs," a word which became distorted through careless speech into "gossip." The godparents had to meet frequently to prepare for the baptism ceremony, and in these meetings much family news was discussed. Such chitchat was called "gossip talk" and finally just "gossip."

We see that almost all English was once slang, some of it dating to Roman and Greek. Respectability of a word for formal usage may be thought of as being on a continuous scale, with very delicate distinctions as to where it can and cannot be used. With each passing decade, more and more slang words make the grade to acceptable speech. The writer of a technical report can make his style more lively and interesting by properly choosing words that are in transition, but he must be careful. If a word is too slangy, it will give a report an atmosphere of flippancy. If one is too timid, however, and uses only standard words, the report may sound pedantic and dull. The right choice and balance is a fine art, requiring an intuitive sense of the standings of words in the mind of the intended reader. A report prepared for a young executive could contain, for example, more recent slang than one written for the perusal of an older, more-conservative reader.

Practice Exercise

Make deductions as to the origins of the following words. Then look them up in the best dictionary you have access to. (The 27-volume "Oxford Dictionary," more properly called "Murray's New English Dictionary," is the most complete; it is found in most libraries.) Also make a guess as to the date at which each word first came into use in its modern meaning. See our answers at the end of this lesson.

1. crane	4. derrick	7. temper	10. fender
2. pistol	5. bloody	8. aspirin	11. galaxy
3. robot	6. watch	9. engineer	12. satellite

Technical Terms

Most *technical* terms are deliberate inventions. For example, the g_m (or *transconductance*) of a vacuum tube refers to the quotient, $\Delta i_p / \Delta e_g$. When the current in any circuit is divided by voltage, the result is conductance. In the case of a vacuum tube, the voltage is at the input and the current is 'across' the tube at the output, so that the prefix 'trans,' meaning across, is appropriate. The word *transistor* was invented by analogy to a resistor. The latter resists the flow of current, while the transistor helps it across from emitter to collector. The derivation of *thermister* as a resistance which varies with temperature is logical. In older times when Latin and Greek were studied by all educated people, new terms were made from these languages. *Photography* meant writing (graphy) by means of light (photo). Telephone meant *distant sound*, while *television* is an unnatural union of Greek and English for *distant vision*.

Men's names are frequently made immortal by becoming common nouns. The electrical units, ampere, ohm, volt, watt, farad, henry, gauss, maxwell, and more recently, hertz, were all once the names of early scientists. The unit of radioactivity, the roentgen, is the name of the discoverer of x-rays. This name was for a time applied to the rays, but Roentgen himself spoiled things by applying the letter, x, referring to the mathematical symbol for an unknown quantity, to the mysterious radiation which at first nobody understood. The expression was so apt and appealing that it became popular at once.

Some names are applied humorously. A unit of area applied to express the projected area of an atomic nucleus which intercepts high speed particles in a cyclotron is *barn*, whose value is 10^{-24} cm². This term is used fairly commonly in scientific papers. It was originally used by Oppenheimer, who said that hitting a nucleus as big as 10^{-24} cm² was like hitting the side of a barn. The origin of this term is forgotten by today's generation of graduate students in physics, who use Oppenheimer's little joke without cracking a smile.

New words usually develop uncontrollably. A current example is *stereo*, applied to a record-playing system using two channels. Such recorded music usually sounds better than monaural music, because the sound presented to each ear is slightly different, duplicating the situation of listening to a live performance. The correct technical term for the process is *binaural* (two-eared), but in the early days of its development somebody thought that there was an analogy between sound recorded by two microphones and the stereoscopic pictures taken by two cameras. The analogy is really poor because the word stereo implies distance to an object, whereas in binaural hearing we sense only the apparent direction of the source of sound. Good or bad, however, the word is with us and the definition of stereo is now: "a record player using two channels."

The Development of English

Language first appeared spontaneously in prehistoric times. We assume that at first each tribe had its own distinct language, but that soon mixing processes came into action. Tribes wandered in search of food, and out of their meetings with other tribes combined languages were born — languages which had bigger vocabularies and were more expressive because they met the needs of larger cultures. Military conquests were even more effective in changing language because the victors usually forced their tongue on the vanquished.

As a result of migration, conquest, and commerce, regions in the world within which travel was not too difficult, developed *language families* which shared words and forms of construction. The sub-continents of India and China became united in similarities among the many local dialects. Regions which enjoyed economic and political stability, like Egypt and Greece, developed distinct languages which influenced those of weaker cultures around them. Barriers like mountain chains and oceans formed the natural boundaries between language families, as they did between empires. The greatest barrier of all, the Atlantic Ocean, resulted in the native American Indian languages being totally different from those developing in Europe and Asia.

The ancestry of English is rooted in several language families. These are the local tongues which developed in the British Isles, the local tongues which developed in the Scandinavian lands to the north, and the Indo-European language family which started in India and is represented by the ancient Sanskrit literature. Through the medium of trade and conquest, the Indian language was imported into Europe where it profoundly influenced Greek and Latin. Latin itself was compounded of native Etruscan and Italian dialects, modified first by Greek and later by the tongues of all the barbarians conquered by Rome. In turn, Latin became the foundation for all the languages within the compass of the Roman Empire — Italian, Spanish, Portuguese, French and English. The first four named are referred to as Romance languages from their Roman origin. The ancient Indian language also influenced Europe directly through bands of wandering gypsies, whom we have already mentioned.

English began as a true native tongue developed spontaneously among the Celts, Picts, and Scots of the islands. Four centuries of Roman occupation consolidated these languages and added Latin words and forms.

With the collapse of Roman power, the British Isles were on their own. The native Britons who were somewhat Romanized were at once attacked by the Picts and Scots from the north. In the year 449, they called in Angles and Saxons from the continent to help repel the invaders. The Angles and Saxons were notorious as pirates along the coast of Flanders, and were early precursors of the Vikings who came four centuries later from farther north.

As frequently happens, the helpers proved to be as bad as the tribes they were called to help against, for the Angles and Saxons took over, forcing Britons to flee to Wales and across the English channel where they formed Brittany. A Briton from Wales fought a successful rearguard action against the Saxons, and gained a place in history as King Arthur.

Whether the conquest of Britain by Angles and Saxons was good or bad politically is, from our standpoint, beside the point. Linguistically, it enriched the embryonic English language with Germanic words and forms, and directly also affected the infant French language as a result of Britons settling on the continent.

About three hundred years after the fall of Rome, England was harassed by Viking raiders from Denmark and the Scandanavian Peninsula. At first the Vikings made only hit-and-run raids, but by 866 AD they commenced making settlements along the coast of the North Sea. For 150 years the Danes and English battled until in 1016 the Dane, Cnut (also spelled, Canute or Knut), became king of all England. (In 1018, Cnut became king of Denmark also, and he reigned as king of

England and Denmark until 1035.) In 1042 Danish rule of England ended with the election of Edward the Confessor, but when Edward died in 1066 the nation fell into confusion between rival claimants for the throne.

The dispute between the English and Danes was settled in 1066 when Duke William of Normandy conquered England, killing King Harold and paying off the Danes. The Norman conquest was the greatest single change in English politics, and it started the greatest reformation in the English language. By 1066 the French language had assumed a form sufficiently close to modern French so that native French speaking people today can understand it. This was the language of William and his nobles, who took over the rule of England. The natives at first hated the French as they did the Normans, but within less than a generation, they were viewing French as the language of culture and status, and taking French words into their speech in order to appear like their new masters. The use of French phrases was considered a mark of elegance, and even today we regard French as somehow more proper than Anglo-Saxon expressions.

English emerged as a modern language several centuries after the Conquest, following a period of adjustment and settling-down. A language may be said to exist when significant literature is written in it. In the case of English, the first significant literature was Chaucer's Canterbury Tales, written in the late 1300's.

Two hundred years later, the place of English as a major literary language was secured by Shakespeare, and the scores of great writers who followed him.

Until the 18th century, English was spoken only by the small population of the British Isles. In contrast, languages like Russian and Arabic were spoken over vast areas of the world. The lucky break for English came with the settlement of North America and the establishment of the United States as an English-speaking nation. The tremendous material growth of America guaranteed the importance of English in science, industry, and commerce, and has resulted in its becoming the most nearly universal language on Earth.

As natives in English, we should feel very fortunate. Natives in other tongues may learn practical English, but they can never have the command of our language that we have. We should follow up this advantage by *mastering* our tongue, and utilizing its marvelous richness in expression. In speaking and writing, we have at our command quotations and phrases from a great literature, and an immense vocabulary that had drawn on every culture in the world. As engineers, English is our major tool in delivering our creations. When we use it fully, we will be both more successful as engineers and as human beings.

Latin and Greek Influence English

We usually learn new words by hearing them used — that is, from context. We can always find official definitions in dictionaries, but these are often not in accord with the delicate distinctions of usage. In order to acquire a full understanding of subtleties of words, we must delve into their history and origins — that is, into the fascinating science of linguistics.

The greatest single source of English words is found in the classic Latin and Greek languages. The study of these tongues today is not undertaken to gain proficiency in speaking or writing Latin or Greek, but for the light such study throws on English words.

Practice Exercise

The importance of Latin and Greek in modern English speech is readily proven by the fact that we can almost read Latin by the similarities of Latin words to English. For example, consider the following Latin words, and find English words derived from them. See our comments at the end of this lesson, after you have written out your answers.

13. barba	18. cumulus	23. dens	28. tacitus
14. corona	19. initium	24. mare	29. notare
15. pluma	20. locus	25. testis	30. palpare
16. rota	21. pallium	26. vis	31. rodere
17. unda	22. cinis	27. plenus	32. torquere

Make similar deductions concerning the following Greek words.

33. ge	40. bios	47. esoteros	54. orthos
34. gonia	41. petros	48. physis	55. moros
35. elektron	42. klima	49. homoios	56. ekklesiazo
36. astron	43. naus	50. stigma	57. schizo
37. atmos	44. pyr	51. gymnos	58. sarkazo
38. lithos	45. phos	52. holos	59. planaomai
39. hippos	46. sperma	53. leukos	60. mimeomai

From the foregoing lists of words, one might be tempted to think that any intelligent English-speaking person should be able to understand Latin and Greek. Unfortunately, there are also many words in these languages which did not migrate into English. These latter include most of the common "little words" describing daily life. For example, we might guess that *baino* would mean *bath*, since this is *bain* in French and *bano* in Spanish, but in Greek it is the verb "to go". Some Greek words have been applied in a specialized way in the arts, like *charasso*, to engrave, which is the root for *charassma*, a word descriptive of the quality of brushwork on a painting.

Many English words have changed so much through the centuries that their Latin and Greek origins are well hidden. For example, *enemy* comes from Latin *in* (not) and *amicus* (friendly). To *scamper* originally meant to get out of the field, ex-campus, and referred to a defeated army. *Companion* is a person who breaks bread with someone, com- (with) and *panis* (bread).

The word *cannon* in English comes from Greek *kanna* (a reed) which is of tubular shape like a cannon. The Italian city of Naples came from *neos* (new) and *polis* (city). The Greek *kathedra* (chair) gave rise to a number of English words: *cathedral* (the seat of a bishop), *chair*, and *shay*. The English word *metaphysics* came about because in Aristotle's compilation of human knowledge his discussion of abstract philosophy came after his discussion of nature, hence *meta* (after) *physika* (physics). The Greek word *damao* (tame) provided the base for *adamas* (untamable) which became the English *adamant*. A Latin form was *adamantis*, which, after dropping the a, gave *diamond*, a material noted for its resistance to change.

We use a number of Latin and Greek expressions directly in common writing. The letters, e.g., stand for Latin *examplum gratias* — and mean in English, *for example*. *Ad hoc* refers to something done for one purpose only, as an "ad hoc committee" which dissolves after doing one job.

Latin plurals are not formed by adding *s* as in English, and this causes considerable difficulty. Latin words ending in *s*, like *hippopotamus*, may end with *i* (hippopotami), but in other cases usage in English has abandoned the Latin ending; the plural of *caucus* is caucuses, not cauci. The word *data*, referring to numerical information, is really the plural of *datum*. Hence it is correct to say "these data" and, if only one number is referred to, "this datum". Unfortunately *datum* has been preempted in technical use to mean a point of reference, so we are stuck with one word only for information. Most engineers ignore the niceties and say "this data", but this still brings on a slight feeling of inelegance among educated engineers, so that we must exercise care. One solution is to use the word in a way that does not disclose whether it is singular or plural — "the following data", "this piece of data", etc. Or one can take the coward's way out and substitute *information* for the sticky word.

Vacuum is another word with a troublesome plural, *vacua*. Most technical works say *"vacuums"* but the spelling is still a bit of a toss-up, and in deciding one might consider the age and erudition of the reader of a report. A vacuum cleaner salesman could be quite annoyed by insistence that he is really selling "Hoover Vacua".

Words can change, not only in meaning but also in social acceptability. In Latin, the word *arlotus* means a glutton. This became *arlotto* and implied drunkenness ás well as overeating. In medieval French it was *arlotte*, and meant a generally debauched rounder and man-about-town. Coming into English, the same general meaning was kept but with variants — meaning a man-servant or valet, and a form of pointed shoes popular in the 1300's among young men. Later the meaning changed to moral looseness, and with the addition of an *h*, it became the modern *harlot*.

In past ages it was a great insult to call one a dog (and it still is, in Arabic countries). Whenever a word gets a bad reputation, people tend to substitute other words or phrases which imply the bad word without saying it. This process was applied in the case of *dog*, and a rather elaborate expression was cooked up to get around the dirty word. Unfortunately, things backfired, so to speak, because the substitute got such a bad reputation that today it is abbreviated as *s. o. b.* (The late Senator Huey Long once received a letter addressed to "Huey Long, s.o.b., Washingon, D.C." A colleague explained that the letters meant Senate Office Building. Replied Huey: "That's not what *my* constituents mean.")

Styles of English

Shakespeare is reputed to have had a vocabulary of 25,000 words. Today educated men know several times this many, but not all of the words are commonly used. The spoken vocabulary is smaller than that used in writing because there is not time to reflect and search the memory for the precise word. An English linguistic expert, C. K. Ogden, once compiled a list of 980 words which he called Basic English, with which one could carry on intelligent conversation. (In Basic English, word-book is used for dictionary, for example.)

In speaking we also tend to use simpler arrangements of words and phrases. Sentences are shorter than in writing, and there are few occasions where transcribed dialog would require a colon or semicolon. Amateur playwrights sometimes get into trouble by using a *written-style* for speech. The result sounds stilted and artificial.

In writing we also use different vocabularies and styles at different times, depending on the subject matter, the degree of formality desired, and the educational background of the intended reader. In later lessons of this Series we shall consider the various styles of writing used in technical reports and letters. Such writings constitute only a part of the larger field of English composition. We can survey this great field briefly by dividing it into the following broad categories:

1. EXPOSITORY WRITING

This type of English composition is intended to convey information, or to explain a viewpoint. Exposition ranges from a technical manual which tells how to service and repair a piece of electronic equipment, to an essay on philosophy. Textbooks are examples of exposition, as is this lesson. The purpose of exposition is to give the reader certain information or ideas, and it should seek to accomplish this goal without unnecessary padding or flourishes. In the course of these lessons, we will give many examples of good and bad expository (ek-SPAHZ-ih-tory) writing. Exposition should above all be clear and well-organized. It should seek to build new ideas on the assumed basic knowledge of the reader, and carry his understanding forward by a definite amount.

The style of exposition should be simple and understandable. It should make pleasant reading by avoiding repetitions of words, and awkward and ungrammatical construction. It should avoid expressions that are merely cute or picturesque. For example, most of the witty expressions quoted earlier in this lesson would be out of place and annoying in a serious technical report. The object of exposition is to teach, and this fact should never be forgotten.

2. NARRATION

Narration is employed in fiction and to some extent in factual writing, as in history. The object of narration is to create the impression of an experience, to produce the same mental images that would be made by travel, by encounters with other people or events, or by adventures in general. A well-written work of fiction in narrative style can take over the whole consciousness of the reader, so that he feels that he is actually living the story presented. When properly used, words are capable of filling the mind with pictures, sounds, smells, the sensations of heat or thirst, and the whole gamut of emotions. The writers of political speeches and advertising are well aware of the potential of narrative English, and they can create a whole scene in a few words which bring up memories common to most people.

Good narrative prose is not easy to produce, but style in this form of English is based on simple exposition. Many great novelists got their basic training as newspaper reporters, just as great artists must begin by studying the elements of drawing and perspective. It is not our object in these lessons to make you into a novelist, but what you learn here should be of help if you decide to try your hand at the great American novel.

3. POETRY

Poetry may be described as either exposition or narration, refined and pruned to the ultimate essence of the idea. It may be regarded as the highest form of English expression, in which the meaning has been concentrated by distilling out all unnecessary ideas. Poetry is usually written with a rhythmic repetition of syllables, and with rhyme. Much modern poetry has abandoned rhyme and rhythm, in favor of progressions of well-chosen words. Some of the most intensely expressive poems are very short, like the little verses of Omar Khayyam, or the Japanese poems which must contain only a certain number of syllables.

Writing of all types is both a craft and an art. It is a craft first, in that the writer must be able to use the tools of his trade — words, grammar, style — before he can even try to express complex ideas. It becomes an art when the writer organizes his thoughts into an effective plan, and presents them in a developing sequence that is clear, easy to follow, and pleasant to read. The art side of writing is clearly vital in poetry and in narrative literature. Although no textbook has yet won the Nobel prize in literature, the writer of expository prose should not feel that it is unnecessary for him to use excellent style. One has but to read the works of the scientist, Sir James Jeans, to see that elegance has a place in technical writing — in both his popular *Mysterious Universe* and his advanced treatise on *Electricity and Magnetism*.

The Technical Writer's Attitude Toward His Work

Good technical writing is a craft and an art, with emphasis rightly placed on the former. For this reason many college English majors who are well prepared to be excellent technical writers feel that technical writing is hack work, and prefer to go into advertising or newspaper writing once they are convinced that they are not about to produce a best-selling novel right away. In such a decision they often miss a challenging career, for technical writing is definitely not hack work; it is far more exacting than the writing of news stories, for example. The technical writer is not a solitary creator in an ivory tower, but a team member whose typewriter is at the very center of activity in industry. Whether you write reports as part of your job of technical development, or devote full time to the writing of manuals, advertisements, proposals, or other material for many engineers, you will find that above all, technical writing is not dull.

Technical Writing

Technical writing as a distinct and recognized branch of English composition is comparatively new. During World War II it received a great impetus thru the need for manuals for military personnel who had to operate and maintain equipment much more complex than that used in any earlier war. Since 1945 the rapid development of technology has produced what is termed an "information explosion". Written material forms a big part of this information.

What is technical writing, and can it be classified as to types? This is not an easy question to answer because of the rapid change inherent in technology, but perhaps the following divisions are reasonable:

1. TECHNICAL MANUALS

These are handbooks (*manual* is from Latin *manus*, the hand) each of which describes the operation, maintenance, or repair of some device or system. They are written for users or servicemen, and each such manual should match the user's requirements in technical content and style. The owner's manual for an automobile, for example, should avoid technical terms and assume zero mechanical knowledge on the part of the reader. Manuals for technicians or technically trained users of equipment may assume basic knowledge — e.g. Ohm's law and perhaps AC circuit theory, in a manual intended for use by electrical technicians.

2. TECHNICAL REPORTS

The writing of reports is a hated chore for most engineers, and the reports they write with such an attitude make dull reading, equally disliked by the executives who must read them. Technical reports are written for a small audience whose interests and technical backgrounds are usually well known. You can write a better, more interesting report if you learn to enjoy writing.

3. TECHNICAL PROPOSALS

Many companies do most of their business in the field of research and development. Their customer (who is often a governmental agency) will award a contract on the basis of a proposal written in the name of the company. A proposal is a selling device. It sets forth what the company offers to do for a specified price, and outlines the general method of solution for problems likely to be encountered. In this last respect the proposal writer must exercise care in limiting the amount of information disclosed. If he makes the solution too easy, the customer may think the price is too high, and if the contract is awarded and things are not as simple as the proposal implied, the company may be accused of incompetence and the proposal writer fired.

4. TECHNICAL ARTICLES

Technical articles are general descriptions of a product or process, written usually for trade journals or house organs. The readership is large and varied in background, so that the article must be more general and simpler technically than a report. Articles written for major magazines like *Scientific American* often have real literary merit, and are written with great care and skill. Sometimes when proprietary secrets are included in a product (such as a tricky manufacturing process), it is as important to know what to leave out as what to put in.

5. WRITING FOR AUDIO-VISUAL PRODUCTIONS

Slide films and sound motion pictures are increasingly used for presentations to customers, stockholders, sales prospects, and employees. Their purpose can be educational, morale building (for employees), or part of a sales effort. When a film is being planned, the technical writer is usually called in, and he may be asked not only to write narration but to assist in the overall planning of the production.

Editing and Production

There are numerous subsidiary jobs related to technical writing. These include *technical editing*, which is like inspection in a factory. The editor may catch errors in spelling or grammar, and question passages which appear unclear to him, but his main function is to arrange paragraphing, numbers for equations and illustrations, and generally put the writing in the form preferred by the company. Editing is most important in proposals, and least important in articles for general publication since the magazine itself may take care of this matter. Illustrations, curves, and other graphic materials are often needed. This work is done by draftsmen or illustrators (renderers) who must work in cooperation with the writer.

The writer may also be called upon to proofread camera copy or galley proofs when his material is printed, and to assist generally in the follow-up of production. A basic knowledge of printing and other reproduction processes is very helpful in this respect.

The lessons which follow discuss the foregoing aspects of technical writing in detail, with examples and exercises relating to the various kinds of writing.

Practice Exercise

Listed after each of the following numbered words are several *possible* synonyms or definitions of that word. Select in each case the synonym or definition which applies.

61. seismic   a. sixth, b. sudden, c. startling, d. earthquake related

62. celerity   a. weakness, b. purity, c. speed, d. famous

63. coalesce   a. bring together, b. force, c. carbonize, d. resist

64. artisan   a. deep well, b. musician, c. craftsman, d. amateur

65. penultimate   a. extreme, b. last, c. next to last, d. written ultimatum

66. entropy   a. temperature, b. randomness, c. weather, d. interior state

67. portent   a. omen, b. guess, c. movable, d. representation

68. equipotential   a. contour, b. velocity, c. depth, d. quiescence

69. evanescent   a. bubbling, b. humorous, c. sedate, d. temporary

70. quintessence   a. typical, b. fifth, c. unlike, d. exceptional

Answers to Practice Exercises

1.   crane. Originally it meant only the aquatic bird. When the loading machine was first invented, country folk laughingly compared it to the long legged bird. The mechanical meaning dates at least to 1375.

2. pistol. Like the names of many things, this word was derived from the place where it was made — in this case, Pistoia, Italy. The original pistol, however, was not a firearm at all, but an oversized dagger first made long before the use of gunpowder. The handgun was given the name humorously at first; then the word stuck when the original meaning was forgotten.

3. robot. This word was invented by a playwright in the 1930s to mean exactly what it does now — an artificial creature which can perform operations requiring human intelligence, but which lacks human sensitivity.

4. derrick. This word for a hoisting machine was originally the name of a famous hangman of the early 1600 s, and was applied because the hoist resembled his gal-

lows. Derrick's name became famous because he was assigned by Queen Elizabeth the unpopular job of executing the Earl of Essex.

5. bloody. In America we think of this word in a literal way, as meaning covered with blood. In England, it is an obscenity and is used as an adjective or adverb — "it was bloody awful", "he ran bloody fast". No explicit meaning is implied, but the word is never used in polite speech. The word is actually a contraction of an oath, "By our Lady", used by Catholics prior to the time of Henry VIII. In the years following this monarch who established the Anglican church, everything Catholic was abhorred in England, and the Catholic oath was deliberately slurred and telescoped together to form a bad word whose origin is now forgotten by the people who use it.

6. watch. The name of the small timepiece derives from its use in the late middle ages by town watchmen.

7. temper. This word comes from French *tempre*, meaning proportion, and was applied to the proportion of ingredients used in making fine steel. Proper heat was regarded as an ingredient in the process, and a steel which had been improperly heat treated was said to have lost its temper. The application of the phrase to a person suffering from loss of emotional control was first made humorously. Today this ancient slang word for mental instability or anger has become respectable standard speech.

8. aspirin. This word was an artificial trade-name of the Bayer Company in Germany. Through common use, it was applied to any pill made from the same chemical material. During World War I when German patents and trade-marks were confiscated by the U.S. Government, aspirin was declared a common noun and the Bayer Company lost the exclusive right to its use. (The trade-name *kodak*, invented by George Eastman to name his camera, is used by many people as a common noun, but the Eastman company has thus far successfully defended its exclusive right to the word.)

9. engineer. This word comes from a Latin word meaning invention, and it once was applied very broadly to any process creating novel results. For example, when the English mathematician, Briggs, first met John Napier, the inventor of logarithms, he is reported to have said, "By what *engine of wit* did you discover this marvel!" In the 1700's the word engineer was applied to the operator of ingenious mechanisms, but after the invention of the steam engine it was reserved for the manager of such engines, either stationary or on locomotives. The modern use of the word for a professionally trained designer has not yet found universal acceptance among common people, who still refer to the janitor as a "building engineer".

10. fender. This word for a part of a car was first applied to rope fenders used to prevent ships from rubbing against a dock, and was originally called a "defender" for the dock against the ship.

11. galaxy. The Greeks, observing the sky without telescopic aid, compared the wide hazy band of stars to milk, and called it a galaxy from the word "gala" meaning milk. (We still call it the Milky Way, although smog-ridden city dwellers never see it). When it was realized that this band of light was composed of billions of stars comprising a system, astronomers Stewart and Tait used it in 1878 in their book, *Unseen Universe*, to mean any ordered star system. The word was immediately picked up and became standard English.

12. satellite. In Latin, satellite meant a bodyguard or attendant, and the term was applied to such men who surrounded a noble or king. When Kepler, in 1611, realized that the starlike objects near Jupiter revolved about the planet, he named them satellites after the king's attendants. The Russian word *sputnik*, meaning fellow traveler, has a similar origin.

13. barba = beard. Derived words: barbarian and barber.

14. corona = crown. An electric corona discharge appears like a crown of light.

15. pluma = feather. Derived: plumage of birds.

16. rota = wheel. Derived: rotation.

17. unda = wave. Derived: undulate.

18. cumulus = heap. Derived: cumulus cloud, accumulate.

19. initium = beginning. Derived: initiate.

20. locus = place. Derived: location, and locus in the mathematical sense.

21. pallium = cloak or cover. Derived: to palliate (cover up, cure superficially).

22. cinis = ashes. Derived: incinerate.

23. dens = tooth. Derived: dentist.

24. mare = sea. Derived: maritime.

25. testis = witness. Derived: to testify.

26. vis = force. Derived: vise.

27. plenus = full. Derived: plenary session (fully attended) ; plenum (air chamber).

28. tacitus = silent. Derived: taciturn.

29. notare = mark. Derived: notary.

30. palpare = to stroke, feel. Derived: palpable (capable of being felt).

31. rodere = to gnaw. Derived: rodent.

32. torquere = to twist. Derived: torque.

33. ge = earth. Derived: geology (earth study).

34. gonia = angle. Derived: trigonometry.

35. elektron = amber. Derived: electricity (from charge when amber is rubbed).

36. astron = star. Derived: astronomy.

37. atmos = vapor. Derived: atmosphere.

38. lithos = stone. Derived: lithography (printing from stone originally, now metal).

39. hippos = horse. Derived: hippodrome, hippopotamus (water-horse literally).

40. bios = life. Derived: biology.

41. petros = stone. Derived: petrology (study of chemistry of rocks, in geology).

42. klima = slope. Derived: incline.

43. naus = ship. Derived: nautical.

44. pyr = fire. Derived: pyrotechnics, pyromaniac.

45. phos = light. Derived: photon, photometer.

46. sperma = seed. Derived: spermatozoa.

47. esoteros = inner. Derived: esoteric (inside knowledge).

48. physis = nature. Derived: physics.

49. homoios = the same. Derived: homogenized.

50. stigma = brand. Derived: stigma, in the sense of a figurative mark on character.

51. gymnos = naked. Derived: gymnasium, where students often went nude.

52. holos = whole. Derived: the word, *whole*. The new technical term, hologram, meaning a picture of the entire object from all sides, was made up from this.

53. leukos = white. Derived: the disease, leukemia, characterized by an excess of white corpuscles, was deliberately named for this reason.

54. orthos = straight. Derived: orthographic projection, in which perspective effects are "straightened out".

55. moros = dull. Derived: moron, and probably morose, a dull sadness.

56. ekklesiazo = to convene. Derived: ecclesiastic, a church originally being a place where people gathered to rehear religious doctrine and history.

57. schizo = to split. Derived: schizophrenic, and schism.

58. sarkazo = to tear flesh like a dog, bite the lips in anger, or speak bitterly. This word, after much evolution in Greek, is the basis of "sarcasm" in English.

59. planaomai = to wander. Derived: planet, originally observed as a star which wandered during the year among other stars.

60. mimeomai = to imitate. Derived: mime (in acting), and the trade-name mimeograph.

61. In Greek, seismos means earthquake; d.

62. In Latin, celer = swift; c.

63. In Latin coalescere means to unite; a.

64. A craftsman, skilled workman. Italian artigiano; c.

65. Next to last; composed of Latin *paene* (almost) + *ultima* (last); c.

66. A thermodynamic measure of unavailable heat energy, which is dissipated from directed motion to random molecular motion; b.

67. Omen, from Latin *portendere* (to stretch ahead); a.

68. An equipotential is a line or surface of constant potential, such as a contour

which represents points of equal elevation or gravitational potential energy; a.

69. Temporary, from Latin *evanescere* (to vanish like vapor); d.

70. Typical. Essence means the characteristic quality of something, while quintessence means typical in the extreme — the very essence of the essence. a.

TEST TC-2

TRUE-FALSE QUESTIONS

1. The technical writer should not be concerned with literary style............. _____

2. We think in words as well as visual images................................ _____

3. The vocabulary used by a technical writer should be superior to that of the intended reader, to show the writer's knowledge............................ _____

4. A writer should misspell and be ungrammatical if he is writing for people of limited education... _____

5. It is estimated that Shakespeare had a vocabulary of about a million words..... _____

6. The prolific use of witty sayings like "black as the ace of spades" usually makes written material more interesting................................... _____

7. The Norman conquest resulted in the greatest single change in the structure of the English language...................................... _____

8. Manuals, reports, proposals, and articles constitute the majority of technical writing.. _____

9. The slang word "gyp" is derived from Egyptian............................ _____

10. The word, camera, is directly related to the word chamber. (Use your dictionary.) _____

MULTIPLE-CHOICE QUESTIONS

11. The word, esoteric, is derived from
    1. Egyptian   3. Danish   5. Latin
    2. Greek   4. Angle-Saxon   _____

12. The opinion expressed in the lesson is that technical writing is mostly
    1. Hack work   3. An interesting craft   5. Very important but unrewarding
    2. Art   4. Dull and boring   to the writer   _____

13. The word, aperture, means
    1. Appearance   3. Opening   5. Cornice of a building
    2. Appetizer   4. Top of a tower   _____

14. King Arthur was
    1. A Briton living in Wales   4. A Viking chief
    2. One of William the Conquerors followers   5. The last of the Anglo-
    3. One of the Crusaders   Saxon kings

15. In general, it is considered possible to carry on simple conversation in English with a minimum vocabulary of about
    1. 200 words   3. 1000 words   5. 25,000 words
    2. 500 words   4. 5,000 words
    _____

16. The word, ambient, means
    1. Surrounding   3. Both right and left handed   5. Soft
    2. Walking   4. Warm to the touch
    _____

17. The word, conduit, means
    1. Behavior   2. Pipe or channel   3. Fruit   4. A plan of action   _____

18. The word, archetype, means
    1. A flirt   4. A Greek orthodox church dignitary
    2. The original pattern   5. An extreme example
    3. A style of bridge
    _____

19. The word, preamble, means
    1. A leisurely stroll   4. An introductory section
    2. A horse's gait   5. The first efforts at walking of an infant
    3. The entrance to a building
    _____

20. The word, technical, comes from
    1. A Latin term meaning to touch
    2. A Greek word meaning artistic or skillful
    3. A German word meaning something manufactured
    4. A Danish word meaning made by hand
    5. An Anglo-Saxon word meaning difficult
    _____

OTHER QUESTIONS

21. Suppose the following paragraph appeared in an article directed toward a non-electronics oriented group of high-school freshmen. Would you rate the article as good, fair, or poor? Give the reasons for your rating.

   Today, the word "transistor" is used by many persons in referring to portable, pocket-size radios. Actually, "transistor" refers to the small, button-like devices inside the radio, and not to the radio itself. The transistors are used to amplify the minute signals (often as small as a few microvolts) picked up by the antenna, to the point where they can be used to drive a loudspeaker.

22. If the same paragraph (above) appeared in an article published in a college newspaper, how would you rate it? Again, give your reasons.

23. How do you rate your current knowledge and skill in the proper use of:
    (1) Spoken English ?
    (2) Written English ?

These lessons were originally written as part of a correspondence course in Electronics Engineering. In that course, the lessons in *technical communication* are not presented consecutively but, instead, are interspersed among other lessons in electronics, mathematics, physics, computer science, etc. For the purpose of teaching some of the fundamentals of technical communication, the sequence of lessons presented in this book is independent of the correspondence course, and the reader may ignore references to the Grantham Engineering Series or "the Series."

LESSON TC-3
The Parts and Construction of Language

In the previous lesson we noted that writing is both an art and a craft, and that proficiency in the craft must precede creative success in the art. This lesson is concerned with the *craft* aspects of writing. Under such a heading we of necessity must face up to some of the less pleasant aspects of high school English courses, like grammar and spelling. These matters need not be the agony that they become in some schools, and we hope that in the presentation which follows, you will discover that the mechanics of your mother tongue is interesting, and that the study of the inner workings of English can be as much fun as similar study of technical subjects.

All languages are first learned by natives in a purely mechanical way by imitation of parents and friends. Infants know no rules of grammar, but slowly develop the feeling that some ways of speaking sound "right" while others do not. The ways that sound right are those used by older people around them. When parents use incorrect grammar, the children follow suit, and patterns of speech developed in infancy are very difficult to change later. Parents who indulge in baby talk and sloppy English do their children much harm in this respect, and the poor English is passed on from generation to generation. The difficulty is often compounded by expressions that children pick up from television, which has perpetuated a variety of pseudo western and gangster dialects — dialects that are not authentic but which reflect the notions of Hollywood writers who would be terrified to meet a real badman.

In this lesson we shall review briefly the basic grammatical construction of the English language, and indicate by numerous examples (good and bad) ways of expressing ideas, with emphasis on common grammatical problems. The best way to improve one's fundamental skill in handling the language, however, is found in reading good examples of its use, and in constant critical evaluation of one's own writings. This lesson will serve as a starter. Your continued progress, however, depends on you.

Parts of Speech

The English language is composed of words. There are several different kinds of words which have various functions in constructing the basic unit of communication, the *sentence*. A sentence is a statement about an action, that is, about something that is done. The action is described by a kind of word called a *verb*. The person or thing responsible for the action is called the subject of the sentence and is described by a *noun* or a *pronoun*. Often (but not always) there is another noun or pronoun representing a person or thing acted upon, which is called the object of the sentence. In addition, there are words called *adjectives* which explain the nature of the nouns and pronouns, and words called *adverbs* which do the same for the verbs. Finally there are several kinds of words which act as a kind of glue in holding sentences together, such as the *conjunctions, prepositions, articles,* etc.

The various kinds of words just mentioned are called the *parts of speech,* and the study of their relationships within sentences is called *grammar*. There are eight parts of speech, as follows:

1. NOUN

A noun is the name of something. It can be the literal name of a person or thing, such as Mr. Jones, Arthur, New York City, or North America. Such true names are called *proper nouns,* and are spelled with capital first letters. Also, a noun can be a general term identifying an object, place, quality, or an action. Examples: dog, river, pencil, happiness, laughter. Nouns of this kind are called *common nouns.* Common nouns are called *concrete* if they name something that we can see or feel, such as a tree or the wind. They are called *abstract* when they name qualities that do not exist in a material way, such as *love, honesty,* or *terror.*

2. PRONOUN

When something must be mentioned a number of times, the repetition of its noun becomes tedious, and a stand-in called a pronoun is used. Examples of pronouns are *I, he, she, we, they, it.*

3. VERB

Verbs express action, or state of being. Examples of action verbs are: run, talk, calculate, accelerate, etc. The state-of-being verb is: *to be.*

4. ADJECTIVE

Adjectives describe, limit, or otherwise specify the nature of nouns and pronouns. Examples: a *tall* man, a *magnificent* church, a *small* quantity, an *infinite* distance.

5. ADVERB

Adverbs have the same relation to verbs that adjectives have to nouns: He thought *quickly;* the sun set *slowly.* An adverb may be used also to further modify the effect of an adjective or another adverb. Examples: a *surprisingly* fast horse (where surprisingly is an adverb modifying the adjective, fast); he ran very quickly (where very is an adverb modifying another adverb, quickly).

6. PREPOSITION

A preposition is a connective word which shows the relationship between a noun or pronoun and another word, usually another noun. Examples: cart *before* the horse, dog *in* the manger, sunset *over* the ocean.

7. CONJUNCTION

A conjunction is a connective word which joins two words, or two groups of words. Examples: dogs *and* cats; elephants are big, *but* antelopes run faster.

8. INTERJECTION

An interjection is a word or phrase interpolated into a sentence or standing by itself, which expresses an emotional state. Examples: Ouch! Hallelujah! What the heck! Gosh!

The Sentence

A sentence is an organized group of words which identifies a *subject,* i.e. a person or thing to speak about, and a *predicate,* which says something about the subject. Both subject and predicate must exist to form a complete sentence. For example, the phrase, "The Bengal tiger," may serve as a subject, but by itself it is not a sentence. We can add adjectives and other descriptive phrases, such as in, "The mighty Bengal tiger, silent king of the jungle". But this is still not a sentence because there is no predicate. We can complete a sentence with a single verb, *hunted,* or we can add a longer phrase, *hunted throughout the night.* The formation of partial sentences, usually subjects alone, is a common error in careless writing. People who write slowly by longhand may make such errors because it takes so long to put down the thought that they may forget what they really started out to say. For example: "The king, upon discovering that the queen had been unfaithful again, and in the royal palace, to boot, which was in itself the supreme insult since ordinarily members of the royal guard, not to mention jealous nobles, were available in ample numbers and with every reason to curry royal favor by reporting immediately such indiscretions." This collection of words is ample and tells us quite a bit about the situation in the palace, *but it is not a sentence.* It has no predicate; it does not say what the king did after making his painful discovery. It can be made into a sentence by the addition of a single verb — e.g., "cursed". While technically correct, such a sentence is horribly unbalanced; by the time we get to the end of all the details we have forgotten the subject, and are left to wonder who cursed — nobles, guards, or even the queen?

Sentences may be classified as having one of three purposes: to state a fact, to ask a question, or to issue a command.

A sentence which states a fact is called a *declarative sentence.* For example, "The president called a meeting."

A sentence that asks a question is called an *interrogative sentence.* "When will the president's meeting be?"

A sentence that expresses a command is called an *imperative sentence:* "You men attend the meeting!"

PRACTICE EXERCISES

1. Break up the example concerning the king and queen, etc. (given in the previous discussion) into several sentences, which convey the overall idea in more manageable bites.

Indicate subject and predicate in the following sentences:

2. Barking dogs don't bite.

3. The rare occasion of a public appearance by the great man is news.

4. A fool and his money are soon parted.

5. The inclusion of women students in the all-male technical college was justified on the grounds that it made a more balanced social life for all.

Indicate which of the following are complete sentences, and why the incomplete one are incomplete:

6. Nobody escapes taxes.

7. Taxes, which fall on rich and poor but more heavily on the rich.

8. Transistors, which can both amplify and oscillate, are often made of germanium.

9. The Q of a tuned circuit, which is a function of the inductance in henrys and the resistance in ohms and measures the rate of decrement of oscillations when the exciting signal has been removed.

10. The purpose of a slotted line, which is primarily to determine standing wave ratios and, through the use of the Smith chart, to determine terminating impedances on a transmission line, was explained.

How Parts of Speech are Used in Sentences

The same word may act as more than one part of speech depending on how it is used in a sentence. For example, if we say "Speed kills," "speed" is a noun. In the sentence, "Let us speed around the course again," it is a verb. In "The speed maniac died at the wheel," it is an adjective. The following sentence employs the word "round" as five different parts of speech: "Our round world — which I shall round on this trip — spins round on its axis, while making the circle round the sun that causes the round of the seasons." In order, this word was used as follows:

1. As an adjective, modifying the noun, world.

2. As a verb, describing the action of travel.

3. As an adverb, modifying the verb, spins.

4. As a preposition, which indicates the relation between "circle" and "sun."

5. As a noun, naming the combination of the seasons in sequence.

PRACTICE EXERCISE

11. Give the parts of speech of the italicized words in the following nursery rhyme, and explain why each is the part named.

Jack and Jill went up the *hill* to fetch a *pail of water.*

Jack fell down and *broke his crown, and Jill came* tumbling *after.*

(The words not italicized represent grammatical forms not yet discussed.)

The Noun

Let us return to that important part of speech, the noun, for a closer look. As we have already mentioned, there are several ways in which we can classify nouns. One way is into *common* and *proper* nouns. A common noun is a general name, applying to many similar things: dog, pleasure, redness, ride. A proper noun is the specific name of a person or thing: Donald Grantham, Los Angeles, Pacific Ocean, Handbook of Chemistry and Physics. Proper nouns are always capitalized in English (all nouns are in German). The word, God, is capitalized when referring to the Christian Deity, but not in the case of gods in general, such as Roman deities.

Nouns are also classified as *concrete* or *abstract,* as we have already mentioned. Concrete nouns have physical reality, either material (as wood, or cat) or as manifestations of physical reality (force, energy, field strength). Nouns characterizing particular kinds of things are also concrete (beggar, revolutionary, hero). Abstract nouns describe concepts which lack physical reality (courage, poverty, democracy).

There is a particular type of noun which refers to a group of objects. One might think that the word "group" itself would be sufficient, and in the highly pruned vocabulary of Basic English, it is the only such noun given. Usage, however, has resulted in the development of a considerable number of special names for groups of objects, particularly of common animals. For example, the following *collective nouns* are so universally used that a wrong application stands out at once either as an error, or as a deliberate figurative expression:

herd of cows, antelope, bison, seals — herbiverous animals in general;

pack of wolves, dogs, or other common carnivores, with certain exceptions;

flock of birds, sheep, chickens;

assembly of people, legislators, or parts of a machine or system;

congregation of people when meeting for religious observances;

bevy of certain birds, usually small — e.g., quail;

company of soldiers, also baboons!

galaxy of stars, also figuratively, of movie stars;

swarm of bees or locusts; *school* of fish; *tribe* of natives;

mob of looters; *gang* of hoodlums; *pride* of lions;

pod of whales; *array* of numbers; *volley* of bullets;

pencil of rays; *clutch* of eggs (when under a setting hen);

wisp of vapor; *drift* of snow; *shock* of hair.

Some collective nouns are so rare as to be incomprehensible to most people, e.g., *fardel, coacervation, conglobation.*

Usually collective nouns are restricted to their related subjects, but sometimes good figurative effect can be produced by misapplying a collective noun. The expression, "a flock of clubwomen," suggests cackling, hen-like behavior; "volley of words" implies rapid-fire speech, while the word "spate" which is properly applied to a torrent of flood-water, is often used with "words" to imply an uncontrolled flood. many arresting and picturesque expressions can be invented by using a collective noun whose subject has characteristics that we wish to attribute to some other subject. For example, "herd" implies senseless following, while "drove" suggests blind obedience. When we say "herd of voters" we attribute lack of independent thought, while "drove of refugees" bring up a picture of helpeless flight before a pursuer.

Advertising writers, and the writers of political speeches are well aware of the images that can be created by the misapplication of collective nouns, and while misapplication is not usually desired in a sober technical report, a little spice of this kind can be added now and then to relieve the tedium.

Collective nouns may be singular or plural, depending on whether the individual members of the group act together or independently. For example, if the members of a jury agreed, we would say "The jury was united in its verdict." If unanimity was lacking, it would be "The jury were arguing over a verdict." The latter expression, although correct, sounds odd, so that it is better to change it to "The jury members were arguing..." Usually the use of a collective noun implies collective action, so that in most cases the collective noun is singular.

Nouns are inflected, or modified in spelling, in three ways: by *number* (singular or plural), *gender* (masculine or feminine), and *case* (nominative, possessive, etc.).

Most nouns form their plurals by adding *s* or *es*. Many nouns taken from other languages, retain the foreign plurals, such as Latin *addendum* whose plural is *addenda,* or *ox*, which has the German plural, *oxen.* Many rules have been set up in an attempt to cover all the ways for forming plurals, but the variation and inconsistency of English has defeated the best efforts of scholars and of frustrated foreigners trying to master it. We as natives must ultimately fall back on what we learned as children, word by word, and form plurals in the ways that "sound right." Some plurals are in the process of change. *Data* is now used as either singular or plural. The correct plural of cow is *kine*, but this has been almost completely replaced by *cows*, while the plural, *brethren*, is replaced by *brothers*, except for members of a religious sect.

There are many compound nouns in which the *s* for plural may be on either one or the other of the two nouns involved. Mothers-in-law and courts-martial are accepted, but the plural of spoonful which logically should be spoon*s*ful , is usually taken as spoonful*s*.

Technical terms of foreign origin tend to keep their original plurals more than common words. Addendum and erratum use the Latin *a* ending to form the plural, while words of Greek origin like analysis, basis, crisis, use the *es* ending for their plurals. Many words of French origin use either Angelicized plurals or French plurals, depending on the social pretense of a speaker. The plural of *beau* can be *beaus* or *beaux*. A bride may discuss *trousseaus,* but on the engraved invitation it is *trousseaux*. The plural of *Monsieur* presents a problem unless we know that it means my lord, so that the French possessive form *Messieurs* (my lords) can be used. *Madam* falls in the same category, only worse. Society pages always say *Mesdames*, while *Madams* suggests supervisory personnel in a very old professional field.

The *gender of nouns* is much simpler in English than in many other languages. *Masculine gender* is applied to male humans, and to certain male animals — generally those for which special names are used for the males, such as cock, bull and stallion. *Feminine gender* applies to female humans, to certain female animals, and to a few inanimate objects, such as ships. *Neuter gender* applies to objects without sex, as stone, tree, boiler, and to most animals and often to children. For example, "The dog barks when it hears a sound", or "The baby cries when it is wet". *Common gender* applies to humans or animals that may be of either sex: bird, parent, cousin, child.

Gender is expressed in English by a change in the noun, or by the use of the word female before a noun. The names for the common animals, particularly the domesticated ones, have masculine and feminine forms. For example, the following pairs represent masculine-feminine: buck-doe, bull-cow, cock-hen, colt-filly, lord-lady, dog-bitch, fox-vixen, actor-actress, hero-heroine. The obscene use of bitch has made most people avoid this word, in favor of "female dog", which is like calling a cow a female bull!

Some words use two plurals, to indicate different meanings. For example, *die* means either a forming tool in a machine shop, or a small cube used in gambling. The plural in the former meaning is *dies,* while in the latter case it is the familiar *dice. Genius* in the plural is either *geniuses,* which are very intelligent people, or *genii,* which are supernatural spirits. But to avoid confusion, *genii* is usually also used for the singular.

Case of a noun indicates the *grammatical relation* of the noun to other words in a sentence.

The *nominative case* refers to a noun that is the *subject* of a verb. In "The dog barked", *dog* is nominative. In less simple sentences, we can find the subject of the verb by asking who, or what, the verb refers to. For example, in "The bear went over the mountain" we ask who went over? The answer is "the bear," which is thus the subject of the verb, went.

The *possessive case* of a noun indicates ownership, origin, manufacture or other casual relationships. It is indicated by adding *'s* after the noun, or if the non-possessive form of the word already ends in *s*, the noun is made possessive by adding only the apostrophe. Examples:

Ownership: Smith's farm; Adams' car.

Source: Mrs. Jones' son; God's country.

Manufacture: Campbell's soup; Shakespeare's plays.

Other association: a week's delay; a day's pay; at sword's point.

The possessive of compound words is formed by adding *'s*, or *'* to the last part of the compound word: mother-in-law's virtues; the Y.W.C.A.'s summer camp; Gilbert and Sullivan's Mikado.

The *'s* form of the possessive is not usually used for inanimate objects. Thus we say "the chapters of a book" rather than "the book's chapters". There are some exceptions to this, however: day's work; month's end; earth's surface; razor's edge. These word-pairs have become so common that they are regarded essentially as single words.

A noun is in the *objective case* when it is acted upon by a verb. In "The dog bit the postman", *postman* is in the objective case. The object can be found by asking whom or what after the verb. For example, in "He did his good deed for the day", if we ask "did what?", it is clear that "good deed" is the object.

A verb may also have an *indirect object.* In "The professor taught the class electronics", the *direct object* is *electronics,* while *class* is the *indirect object.*

Besides being the object of a verb, a noun can be the object of a preposition, as: the shade *of* a *tree;* a drop *in* the *bucket;* snow *on* the *ground;* Stanford *versus* *UCLA.* The objects in the foregoing are tree, bucket, ground and UCLA.

A special kind of objective case is called *apposition,* in which a noun is used to provide further identification. "They promoted *Jack,* the Mayor's *son*"; "They arrested the *burglar, Baby Face Floyd." Son* and the name, *Baby Face Floyd,* are called *appositive objective nouns*. Nouns which are *in apposition* explain each other.

PRACTICE EXERCISES

12. Certain quantities or units of measure are commonly associated with particular commodities. Name the thing that first comes to your mind associated with the following *measures:* a. dozen, b. quart, c. gallon, d. ream, e. finger.

13. Which of the following nouns are collective? a. committee, b. books, c. Germans, d. people, e. classics, f. assembly, g. officers, h. ministry.

14. Give the correct plural for each of the following foreign nouns: a. alumna, b. alumnus, c. phenomenon, d. thesis, e. medium, f. gymnasium, g. tableau, h. bandit, i. dilettante, j. libretto.

15. Give the singular forms of the following plural nouns: a. alms, b. dregs, c. pants, d. scissors, e. billiards, f. blues.

16. Give a noun which fits each of the following descriptions: a. a man who owns and rents houses, b. a lady who can see into the future and tell fortunes, c. a male cat, d. the wife of a baron, e. an old maid, f. the equivalent of *mister* in Spanish, Italian, Portuguese, Japanese, g. a young female horse.

The Pronoun

A pronoun [Latin *pro* (for) + *nomen* (name)] is a substitute for a noun, and is used to prevent repetition and awkwardness. A pronoun should be identified with a noun (called its antecedent) in the same sentence, or in an immediately preceding sentence. When there may be doubt as to which antecedent applies, then it is best to repeat the noun. Consider the following examples:

1. Ed missed the bird because he was blinded by the sun. (The proper noun, *Ed*, is the antecedent of the pronoun, *he*.)

2. The hunter waited two hours at the waterhole. Finally he went to sleep. (*Hunter* is the antecedent of *he*. There is no confusion since only one noun is involved.)

3. The captain ordered the mate to fight the fire. After an hour he abandoned the ship. (It is not clear whether *he* refers to *captain* or *mate*, since it seems equally likely that either one might abandon ship. Here one should use a noun instead of the pronoun, *he*.)

There are several kinds of pronouns, as listed and discussed below.

1. PERSONAL PRONOUNS

These pronouns refer to the person speaking *(first person)*, the person spoken to *(second person)*, or the person spoken about *(third person)*. Each of these persons has three cases: nominative, possessive, and objective. In the first person these are *I*, *my (mine)*, and *me* in the singular, and *we*, *our (ours)* and *us*, in the plural. There is sometimes confusion over the use of *I* or *me* by some people, in cases such as, "James and I went fishing". When in doubt, the trouble is resolved by leaving out the "James and" and checking whether *I* or *me* sounds right. Among the uneducated, expressions such as "Him and me were late" are common, in which for both pronouns the objective case is used instead of the correct nominative, *He* and *I*.

2. DEMONSTRATIVE PRONOUNS

These pronouns serve to *point out* a person or thing. They are *this, that, these,* and *those.* For example, "This is where I live", "Those are the fresh eggs".

If a demonstrative pronoun is used with the noun it refers to, it becomes an adjective, as in the following: "This house is mine" or "Those eggs are fresh."

3. INDEFINITE PRONOUNS

An indefinite pronoun is a general stand-in for one or more nouns. For example, "You don't know *everything*", "*Somebody* must know", "*All* must suffer". Typical commonly-used indefinite pronouns are the following: *all, another, any, anyone, anything, both, either, few, least, may, much, none, nought, one, several, some, something.*

4. RELATIVE PRONOUNS

These pronouns have two functions. A relative pronoun replaces a noun and acts as a connective bridge between parts of a sentence. For example, "It was a noise *that* could be heard for miles", "This is the man *who* said it." The relative pronouns *which, that,* and *what* are the same in nominative and objective cases, and add "of" for the possessive. The pronoun, *who,* changes with case, as shown in the following examples:

Nominative: This is the man *who* owns the dog.

Possessive: This is the man *whose* dog barked.

Objective: This is the man to *whom* I spoke.

Confusion often occurs in deciding whether to say who or whom. If the pronoun denotes the person acting, it is *who*; if it is the person acted on, it is *whom*. If the pronoun is preceded by *to* or *for*, it must be *whom:* "To whom shall I speak", or "For Whom the Bell Tolls".

The pronoun, *which,* refers only to things or animals, not to humans. E.g., the *resistor* which overheated, the *calf* which bawled. *That* refers to either humans or non-humans: "The man *that* sings is happy", "the capacitor *that* shorted out".

Compound relative pronouns are formed by adding *ever,* and *soever* to *who, which,* and *what.* E.g., *whoever, whosoever, whichever,* etc. Some of these are very commonly used: "Whatever have I done with it?"; "Whoever speaks now will be a marked man". Such a compound pronoun as "of whichsoever" may make a sentence awkward, and is usually best avoided.

5. INTERROGATIVE PRONOUNS

These pronouns are used in asking a question: "*Who* will help me?" "*Which* of you will be first?" "*What* amplifier is that?"

6. NUMERICAL PRONOUNS

These pronouns are numbers standing for nouns, as in "every *tenth* (subassembly) will be tested"; "*two* (of the systems) failed the first day."

7. REFLEXIVE PRONOUNS

These are formed by adding *self* to a personal pronoun, as in "he thinks of *himself* first"; "they yelled *themselves* hoarse".

8. RECIPROCAL PRONOUNS

This construction represents interchanging action, as "they pounded *each other* on the back"; "they pounded *one another*".

SUMMARY ON THE CORRECT USE OF PRONOUNS

Misuse of pronouns constitutes a large part of the grammatical error which makes writing sloppy and confusing, and creates an image of an ignorant or uneducated person for the writer. The correct use of pronouns can be summarized in a few simple rules:

Pronouns must *agree* with their antecedent nouns in number, person and gender. Examples: "Henry gave *his* consent"; "Henry and Sam gave *their* consent". Certain nouns which are singular give an impression of plurality, so that a plural pronoun is often *incorrectly* used. For example, "Everybody gets food in *their* turn" is wrong, since "everybody" is singular. The correct form is "Everybody gets food in *his* turn". (If the people involved are known to be all women, *her* may be used instead of *his*. Otherwise, it is correct to use *his*.) The following nouns are *singular*, and require singular pronouns: *each, either, neither, everyone, no one, everybody, nobody*.

Some collective nouns can be either singular or plural, with the choice indicated by the construction. For example, if we say "The jury were divided" we clearly mean jury to be plural. Therefore a following sentence might be *"They* could not reach a verdict", and not *"It* could not reach a verdict". The sentence "Neither Jones nor Smith wanted the dog as *his* own" implies that the dog would belong to one or the other. If *"their own"* is used, it implies joint ownership. Errors of number agreement are frequent in "neither - nor" sentences: "Neither the Republican party nor the Democratic party wanted Smith as *their* candidate" is wrong, since only one party could have him. The correct pronoun is *"its"*.

When a pronoun disagrees in person (first person, second person, third person) with its antecedent noun, the sentence sounds uneven and confusing. For example: "The company insists on high quality; *we* are not perfectionists, but *one* must make parts that function, or else *you* will turn out a worthless product." This sounds uneven because of the disagreement among the pronouns *we, one* and *you*. Such a mess can be salvaged by writing something like this: "We insist on high quality in our company; *we* are not perfectionists, but *we* believe that *we* must make parts to function, or else turn out a worthless product." Technical editors are resigned to making such repair jobs on manuscripts written by otherwise competent engineers.

Agreement in gender is usually done automatically because of our sex conscieousness. We seldom make mistakes like, "Every man will do her duty".

PRACTICE EXERCISES

In each of the following, choose the pronoun which is correct.

17. To the person (who, whom) trusted her she was faithful.

18. She was faithful to (whoever, whomever) she was trusted by.

19. He bit (him, himself) on the lip.

20. All of us were present, Finkelstein, Folowitz and (I, me, myself).

21. The two goats butted (each other, one another).

22. I got the news from the technician (who, whom) knew the engineer to (who, whom) the chief engineer had given the hint.

23. Everybody has a right to (his, their) place in the sun.

24. Between you and (I, me), I think neither (he, him) nor (she, her) knows how to use an ohmmeter.

25. I believe (they, them) to be most competent.

26. (Who, whom) the professors approve are promoted; those (who, whom) are found in the bottom half of the class are not.

27. All but (he, him) had left the shop.

28. Than (who, whom) would you say he is better qualified?

The Verb

Verbs are where the action is. They tell what the subject accomplishes, and what happens to the object. Because of their importance, everybody in history has taken a hand in the care and feeding of verbs, with the result that they constitute one of the most complex areas in grammar. About the only thing we can be thankful for is that English is our language — that it is not one like German, in which the complications are much greater. (No German will agree with this statement, but instead will point out all of the exceptions to rules and irregularities in some English verbs which drive foreigners to distraction.)

Like other parts of speech, verbs can be classified in several ways. One basic distinction is that between *transitive* and *intransitive* verbs. A transitive verb requires an object to act upon, while an intransitive verb does not. In "The dog bit the cat", *bit* is transitive because it requires something to be bitten such as a cat. On the other hand "The dogs barked" uses an intransitive verb *barked*, since it is possible to bark without an object, and dogs often do all night.

Six verbs which are often used wrongly are: *set, raise, lay, sit, rise,* and *lie*. The first three are transitive and require an object: "*Set* the meter on the bench", "*raise* the voltage", *lay* the screwdriver down". The other three similar verbs are intransitive: "he *sits* on the chair", "the voltage gradually *rises*", "the meter *lies* on the bench".

The following *kinds of verbs* have been defined by grammarians:

1. LINKING (OR COPULATIVE) VERBS

This type of verb is intransitive, which acts to join the subject of a sentence to the predicate. Examples: "Happiness *is* a warm puppy"; "All *seemed* lost"; "Our fears *proved* groundless".

2. AUXILIARY VERBS

These verbs modify or assist other verbs in their meaning. For example: "He *could* barely *read* the deflection of the meter needle." *Could* is an auxiliary verb which modifies the effect of *read*. *Barely* is an adverb modifying *could*.

3. STRONG AND WEAK VERBS

The distinction between strong and weak verbs is purely one of spelling. Strong verbs form their *past tenses* by changing a vowel within the word, as in *sing, sang, sung*. Weak verbs require additional letters, as *love, loved, loved*. The names "strong" and "weak" were applied to these two verbs to indicate that the latter did not have the "power" to form past tenses without the assistance of the *ed* endings — a rather curious reason.

Principal Parts of Verbs

In speech it is necessary to distinguish the time of occurrence of the event described. The event may be taking place currently, it may occur in the future, or it may have already occurred. In the latter case, we distinguish between the idea of occurrence anytime in the past, and an event that preceded some other past event. The *verb forms* which make these time-distinctions are called *tenses*. Most verbs indicate their tense by changes in spelling, or the addition of auxiliary words, in a consistent way. These verbs are termed *regular*. Other verbs depart from the regular form in various ways, and so are called *irregular verbs*. The most frequently used verbs tend to be most irregular, possibly because frequent use encourages meddling. To illustrate tense, consider the regular verb, *love*, and the irregular verb, *go*.

Present tense:	I love	I go
Future tense:	I shall love	I shall go
Past tense:	1 loved	I went
Present perfect tense:	I have loved	I have gone
Past perfect tense:	I had loved	I had gone
Future perfect tense:	I shall have loved	I shall have gone

The *future perfect* tense indicates a time in the future preceding some other future event.

In addition to the foregoing tenses, there are verb forms which indicate *condition* on some other event. Examples: "If I can, I *should go*"; "If I had been able, I *should have gone*".

Verbs are also modified to express the state of mind of the speaker. These modifications are called *modes*. Actually, *mood* is more descriptive if we think of it as the mood of the speaker. There are three modes, the *indicative*, the *imperative*, and the *subjunctive*.

The *indicative mode* is a simple statement of fact. Example: "The Earth is round."

The *imperative mode* is used to issue orders or requests: "Stand at attention"; "Look at that sunset"; "Please listen more closely."

The *subjunctive mode* indicates that there is *doubt* in the mind of the speaker, or that the statement is *conditional* on something else. It is also used to describe something imaginary, or wished for. Consider the following examples:

1. If I be here then, I shall gladly help.

2. If I were there, I would gladly help.

The first example, using *be*, sounds stilted and archaic, and is rarely used today. The usual expression is to substitute the ordinary present tense, letting the "if" indicate the conditional nature of the statement: "If I am here, etc." The second example is typical of the way the subjunctive is used. Other forms, like "If I had been there at the time, it wouldn't have happened," or "If I could have been there, etc." are common in cultured speech. In the second example, notice that where *was* would otherwise be used, in the subjunctive mode we use *were* instead.

The subjunctive mode imposes changes in spelling, some of which are rather deliberately ignored because of the general drift away from the subjunctive present tense forms. For example, the correct form is "If he *survive*, he will be crippled for life," but common usage uses the present "If he *survives*, etc." In other situations, the *s* is not added and the correct subjunctive is used: "It is necessary that criminals be punished; it is just that a guilty man suffer."

The rules of grammar tell how to use the subjunctive mode, but they do not detail where and when it be used, or when the subtleties of usage say that its use will interfere with the smooth assimilation of ideas. The only way to sharpen judgment in such matters is to read examples of good writing, to build up an intuitive taste and feeling for the best style in a given situation.

Verbs also have what is called *"voice."* There are two voices, *active* and *passive*, and these are easily illustrated. Active: The damaged meter *gave* a wrong reading. Passive: A wrong reading *was given* by the damaged meter.

The active voice gives emphasis to the agent doing the action, while the passive voice draws attention to the action. In the first example, we are concerned about the damaged meter, while in the second, our attention is given to the wrong reading, with the condition of the meter being noted only to explain why the reading was wrong.

In most cases the active voice is preferred. If the acting agent is not known, or is immaterial, then the passive voice is best. "The meter was damaged." If the person responsible was unknown, the active voice would have to say, "Some unknown person damaged the meter," which is too wordy.

Two other verb forms should be mentioned. The *progressive form* indicates that an action is or was continuing. Where the simple present is "She sings," the progressive is "She is singing." Future progressive is: "She will be singing." The emphatic form uses "do" or "did" to stress that the action is indeed done. "She does sing" or "She did sing" are typical.

A special note should be made with respect to the future tense. The normal future is indicated by "shall" for the first person, and "will" for second and third person. Thus "I shall play the piano" or "she will play the piano" indicates intention. If the "shall" and "will" are interchanged, the meaning is changed to an expression of *determination* to perform the act. "I will play the piano" implies that any obstacle will be surmounted in order to do so, and "she shall play" suggests that either she anticipates using force to do it, or perhaps that she will be forced to play. The exchange of shall and will causes much confusion, and the blind wisdom of usage is forcing the use of "shall" for the first person to indicate determination, in line with its use in the other persons. Unfortunately, purists still insist on the old way, and writing which does not conform may be criticized. In writing this series of lessons we have generally used the old form, except when it sounds too pedantic in context. In that case, the rule is deliberately flaunted!

PRACTICE EXERCISES

Correct errors in the verbs.

29. The prisoner was *hung* at dawn.

30. The shop *loaned* the gauge to Mr. Smith.

31. At the shock, he *sprung* to his feet.

32. The collection of instruments which *are* on the shelf *are* obsolete.

33. There *is* a voltmeter, an ammeter and a wattmeter; now the test can be made.

34. *There's* errors galore in the data.

35. The chief engineer, aided by a picked crew of technicians, *solve* most problems.

One final verb form must be mentioned: the *infinitive*. This is indicated by placing "to" before the present tense in most cases, although there are exceptions. The infinitive form is used to draw attention to a verb as a word. As such, the infinitive acts as a noun. "*To eat* well is always pleasant" uses the infinitive "to eat" as the subject of the sentence. In "I like *to eat*," it is the object of the verb, like.

The Adjective

Adjectives are used *to make nouns or pronouns more specific* — in grammatical terms, to *modify* them. Adjectives are classified according to their functions, as follows:

Descriptive adjectives are either *common* or *proper*. In the common type, they described broad classes of things: "timid mouse," "colorful flowers," "brave soldier," "accurate meter," "high frequency." In the proper type, they are limited in their range of application, to one member of a class. Originally all proper adjectives came from proper nouns, and as such were captilized: Roman holiday, French cuisine, Tesla coil, Panama hat. When proper adjectives have been used for a long time, the origin may be forgotten, and the capitalization omitted, as in: india ink, italic type, galvanic current, pasteurized milk.

Limiting adjectives restrict the application of nouns. Various names are used for these adjectives, as shown in the examples:

Demonstrative: *this* book; *that* dog.

Interrogative: In *which* direction did they go?

Relative: Select *which* meter you like.

Indefinite: *some* days; *every* child; *any* place; *both* sides.

Possessive: *my* slide rule, *your* wrench.

Intensive: The *very* idea.

Identifying: The *same old* complaint.

Numerical: *Three* volts, the *third* attempt.

A special kind of adjective is the article: "The" is the *definite article*, while "a" and "an" are *indefinite articles*. "An" is used with nouns beginning with a vowel, or with a silent "h" — *an* ammeter, *an* hour.

In English, adjectives are placed before the words they modify, with a very few traditional exceptions, such as in "life everlasting" and "time enough."

Adjectives can be applied in varying *degree*.

Normal or positive degree: "For $49 you can buy an *accurate* meter." "Newton was a *great* physicist."

Compartive degree: "A $150 meter is *more accurate* than a $49 meter." "Newton was *greater* than Prof. Jones."

Superlative degree: "The international standard ohm is the *most accurate* standard of resistance." "Newton was the *greatest* physicist who ever lived."

For most adjectives of one syllable, the comparative is formed by adding *er*, while the superlative is formed by adding *est* or *st* to the positive form. For adjectives of more than one syllable, "more" is used for the comparative and "most" for the superlative degree.

Adverbs

Adverbs are to verbs what adjectives are to nouns and pronouns. They *alter or qualify* the meaning of the verb. Simple adverbs answer definite questions which might be asked about the effect of the verb.

Time: The question answered is "when?" Example: The design will be complete *next week.*

Place: The question is "where?" Example: He will come *here.*

Manner: The question is "how?" Example: He speaks *clearly.*

Degree: The question is "how much?" Example: The wire was *very* hot.

Cause: The question is "why?" Example: *Why* does the capacitor discharge?

Conjunctive adverbs act like conjunctions in joining clauses in a sentence. For example, "He measured the resistance; *however,* he failed to note the temperature."

Some of the commonly used conjunctive adverbs are: accordingly, hence, additionally, however, also, anyway, consequently, besides, furthermore, in other words, in short, indeed, moreover, namely, nevertheless, still, on the contrary, therefore, on the other hand, yet, yes, no.

Two short sentences can be combined by any of the foregoing adverbs to make a longer sentence the import of which is largely determined by the relation between the two parts, established by the adverb. Consider two rather unrelated ideas: "He enjoys Beethoven symphonies," and "He is an immoral wretch." All of the above listed adverbs can be inserted between these to create sentences which put forth different ideas — some of which are dillies! Try it!

Adverbs and adjectives are closely related in function. Most adverbs are formed from adjectives by adding *ly* to the latter: swift, *swiftly;* hot, *hotly;* etc. The *ly* ending is not a sure sign of an adverb, however; ugly, for example, is an adjective.

Adverbs, like adjectives, come in three degrees of emphasis: positive, comparative, and superlative. Example: rapidly, more rapidly, and most rapidly. One-syllable adverbs generally don't use the "more" and "most" but take the form: fast, fast*er*, fast*est;* late, lat*er*, lat*est*.

Some adverbs are irregular in the way they form the comparative and superlative forms. Examples: well, better, best; little, less, least; ill, worse, worst. The adjective "far" splits into two forms. For the comparative we have "farther" or "further," and for the superlative, "farthest" or "furthest." Confusion in the use of these words can create a subtle feeling of confusion in writing. The form using the *a* implies distance in a literal sense, as in "Manila is farther than Tokyo," while the *u* spelling refer to degree or quantity as in, "There will be *further* information tomorrow."

The force of an adverb can be lessened by using "less" and "least" before it, although a few by their nature do not admit of any modification: fatally, absolutely, entirely, uniquely. These words, like pregnancy, come in one degree only. The adverb probably most misused in this respect is "unique." If something is unique, there is only one in the universe, and it is ridiculous to say that something else is more or less unique — as the only way this could be true would be to have none of the second at all.

Conjunctions

Conjunctions are a glue which joins words or phrases together. There are two kinds: coordinating conjunctions, and subordinating conjunctions. There are six coordinating conjunctions: and, but, for, not, or, yet, which join words or groups of words. Examples: "cats and dogs," "rain or shine," "simple but elegant," "Beauty fades but character remains."

Subordinating conjunctions attach clauses together. Examples: "I strip wires *while* he solders"; "The clock stopped *because* the battery ran down."

The circumstances in which subordinating conjunctions are used are as follows:

Time: as, often, before, until, while, as long as, as soon as.

Reason: as, because, inasmuch as, since, why.

Condition: although (though), if, unless.

Purpose: so, that, lest, in order that.

Comparison: than.

Simple examples of the foregoing are:

Time: He moved the wires *before* the solder cooled.

(Note that any of the "time" conjunctions could be substituted in the above, although some do not make much sense technically — e.g., until.)

Reason: He quit work, *inasmuch as* the power had failed.

Condition: I'll do it *if* you will.

Purpose: He backed away *lest* the package explode.

Comparison: I earn more *than* you do.

Prepositions

Prepositions also show *relations*. The relation is that *between a noun (or pronoun) and some other word.* Examples: "dog *in* the manger," "moon *over* Miami," "water *under* the bridge." The first noun in these examples is the subject, and the last noun is the object of the preposition. The preposition usually precedes the object, as in the examples, but custom has created many situations in which the preposition follows: "What are we here *for*?"; "What are you talking *about*?" When a preposition is misplaced in a situation not sanctified by usage, it may sound very awkward. Winston Churchill, a master of English prose, composed a famous example of this in his retort to a critic of his grammatical style: "This is the type of arrant pedantry up *with* which I will not put!" — instead of "... which I will not put up *with*."

Prepositions express relations in a number of ways:

Position: the papers were *on* file.

Time: well *before* dawn.

Instrumentality: The tool broke *through* neglect.

Manner: He smiled *with* warmth.

Purpose: We are here *for* work, not *for* play.

Interjections

Interjections are exclamàtions indicating emotion. They have no grammatical connection to adjoining sentences. Typical interjections are: Oh; Well; Ouch; Help; Indeed; Oh dear. Many of the commonest interjections employ four-letter words which unfortunately cannot be cited in this decorous work.

Many interjections originated as appeals to deity, or as oaths — that is, as expressions claiming divine guarantees of veracity. An oath like "Let God be my witness that I speak the truth" in the long form, becomes "By God!" when time presses. Some resounding oaths have been distorted through the years until their original meaning is no longer apparent. For example, "Zounds!" and "Gadzooks!" which TV pirates are supposed to say, were originally oaths taken on holy objects, namely "By God's wounds" and "By God's hooks" — hooks being a synonym for nails used in the 1500's, and in this case referring to the nails with which Christ was affixed to the cross.

Verbals

Nouns and other parts of speech can be derived from verbs. Words so derived are called *verbals*. There are three kinds of verbals: *gerunds, participles*, and *infinitives*.

1. The *gerund* is a noun derived from a verb. The spelling (or form) depends on the tense and whether it is used in active or passive voice. The following tabulation shows how it is done:

TENSE	ACTIVE	PASSIVE
Present:	reading	being read
	writing	being written
Perfect:	having read	having been read
	having written	having been written

A gerund can act in any of the capacities of a noun, subject, object, etc. The sentence, "Marriage is her object" is almost equivalent to "Marrying is her object," using the gerund. To a foreigner, these sentences might appear identical in meaning, but we, as natives, recognize that the former means that she wants to attain the stable condition of marriage, while the latter means that she wants to get the ceremony over with.

2. The *participle* is an adjective derived from a verb. "The *barking* dog" and "the *running* man" use the *ing* to indicate present tense. Other examples: "*Having studied* the manual, he was ready to proceed"; "The neglected wife left home"; "*Having eaten* the bone, the dog slept." (In the last example, bone is the object of the participle.)

When a participle is used, the sentence must make it clear what noun the participle modifies. When there is more than one noun, it is possible to compose sentences without indication of which noun the participle applies to; such participles are referred to as "dangling participles" and are one of the commonest errors in writing, technical and otherwise. For example, "Cursing and screaming, the peace protester was handcuffed by the cop," does not make it plain which one was making the noise. All is clear when we rewrite it as, "The cop handcuffed the cursing and screaming peace protester."

3. The infinitive is usually expressed by placing "to" before the present tense: "to go," "to sing." It can function as a noun, as in "*to err* is human, *to forgive*, divine." It can be used as an adjective as in, "I like beer *to drink*," or as an adverb as in "He stood up *to cheer*."

Many errors can be made with the infinitive. If the "to" is separated from the rest of the infinitive, the reader may lose track of the fact that an infinitive is being used, and think that the "to" is independent, and that the rest is a verb. This situation is called a "split infinitive." If the split isn't too wide, the reader can usually get the idea anyway, but if too many words have been crowded in between, chaos results. Consider the following: "The chief engineer wanted to carefully and with every possible precaution compatible with the safety of life and property measure the voltage." The fourteen words between "to" and "measure" split this infinitive so wide that few readers would get more than a vague annoyance out of the whole effort. The mess can be repaired by changing it to, "The chief engineer wanted to measure the voltage carefully and with every possible . . . etc."

Sometimes a split infinitive is used deliberately and with good effect: "I want you to slowly stir the mixture." In order to avoid the split we would have to say "I want you slowly to stir . . ." or "I want you to stir slowly the mixture," both of which are awkward. In this case, we might say that we have invented a special verb, *to slowly stir*, in which case we do not have a split infinitive at all.

Summary

This has been a long lesson because it seeks to review a very large and complex subject. Were this lesson written for young studnts, it would have to be supplemented by many hundreds of examples and exercises. You, as a mature reader, have been through all of this work previously, and you are presumed to have a practical skill in using correct grammar. If you are like most people, you studied grammar once and then forgot it, and now rely on automatic programming to make you speak correctly. However, at the start of a course specifically devoted to writing in a complex and specialized subject-field, a re-run of the theoretical skeleton that lies inside the skillfully turned phrase is worth while. (Now, *that* was quite a *figure of speech!*)

In any event, you are not expected to remember everything in this lesson, but only to let some of the material rub off as you go through it. Feel free to refer to the lesson in answering questions on the test which follows. In fact, a lot of very good learning is accomplished during tests, by sneaking a peek at the book.

With these two lessons as starters, we are now ready to plunge into the dynamic subject of technical writing.

Answers to Practice Exercises

1. This job can be done in many ways. The following is typical: "The king cursed upon discovering that the queen had been unfaithful again. He was particularly disturbed because it had been done in the royal palace to boot, which was the supreme insult. Ordinarily members of the guard, not to mention jealous nobles, were available in ample numbers, and they had every reason to curry royal favor by reporting immediately such indiscretions."

2. Subject: barking dogs.   Predicate: don't bite.

3. The first 11 words are the subject; "is news" is the predicate.

4. Subject: "A fool and his money". Predicate: "are soon parted".

5. "inclusion" is the basic subject, modified by the other words ending with "college"; "was justified" is the basic predicate, with the modifiers which follow.

6. Complete sentence.

7. Incomplete. This is a subject without a predicate. The long phrase modifies

the word "taxes", but nothing is said about what taxes do. Perhaps "cannot be avoided" might be a suitable predicate to add.

8. A complete sentence. The part before the second comma is the subject, and after this comma is the predicate.

9. Incomplete. It is all subject, the part after the comma being a modifier. This phrase could be changed to a complete sentence by omitting the word, *which*.

10. A complete sentence. It is an awkward one, because of the very long subject, with the predicate, *was explained*, added as though it were an afterthought. This would be better expressed in two sentences as follows: "The purpose of the slotted line was explained. This purpose is primarily to determine standing wave ratios, etc., etc." Note that we should not say, "This is primarily to explain", since it is not clear whether "this" refers to *purpose* or *slotted line*. It is a common error of writers to use the word "this" in reference to something menioned in a previous statement. The writer knows what "this" applies to, but the reader often has difficulty in deciding *which* of several words is meant, and he either has to go back and re-read, or else (more often) just goes on reading with a vague feeling of confusion.

11.	*Word*	*Part of Speech*	*Reason why*
	Jack	proper noun	name of a person
	and	conjunction	joins two nouns
	Jill	proper noun	name of a person
	went	verb	tells what subject did
	up	preposition	gives relation between "went" and "hill"
	hill	noun	names a thing
	pail	noun	names a thing
	of	preposition	gives relation between water and pail
	water	noun	names a thing
	fell	verb	describes action
	down	preposition	gives the relation between Jack and the hill
	broke	verb	tells what happened to Jack's crown
	his	pronoun	stand-in for the proper noun, Jack
	crown	noun	names an anatomical part of Jack
	came	verb	describes Jill's action
	after	preposition	shows the time relationship between Jack and Jill

12. Possible answers: a. eggs, b. milk, c. gasoline, d. paper, e. rum. (Note: when a glass is grasped in the hand with three fingers wrapped around it, the quantity corresponding to the width of the fingers was called "three fingers of rum" by the old Caribbean pirates.)

13. committee, people, assembly, and ministry are collective nouns.

14. a. alumnae, b. alumni, c. phenomena, d. theses, e. media, f. gymnasia, g. tableaux (also tableaus in common usage), h. banditti (bandits), i. dilettanti (dilettantes), j. libretti (librettos seldom used).

15. This one is a catch question, as you have doubtless guessed. These words all have no singular forms in modern usage. *Pants* and *scissors* were once considered to be in two parts; pants evolved from long stockings attached at the top. *Blues* is a slang word invented in that form. *Alms* and *dregs* are thought of as representing several things at a time.

16. a. landlord; b. seeress; c. tomcat; d. baroness; e. spinster; f. Senor, Signor, Senhor, san; g. filly.

17. who; 18. whomever; 19. himself; 20. I; 21. each other; 22. who, whom; 23. his; 24. me, he, she; 25. them; 26. whom, who; 27. him; 28. whom.

29. hanged *(hung* refers to non-human objects, as for example in, "the goose hung high"); 30. lent; 31. sprang; 32. *is* in both cases, instead of *are.* There is only one collection on the shelf and it is the thing that is obsolete. Some of the individual instruments might be current models; it is the collection as a whole that we are speaking of. 33. are; 34. there are; 35. solves.

TEST TC-3

TRUE-FALSE QUESTIONS

1. A noun may be either the subject or the object of a verb...................

2. In formal "correct" English, proper nouns may or may not use capital letters..

3. "Beautiful" is an adjective, but "beautifully" is an adverb...................

4. The noun, "abstraction", is itself an abstract noun.......................

5. The lack of an object does not disqualify a group of words from being classi-
    fied as a sentence...

6. The sentence, "This meeting will be attended by absolutely everyone", is an
    example of an imperative sentence..................................

7. The word, "congregation", is a collective noun........................

8. In the sentence, "The cat was chased by the dog", cat is the subject...........

9. A gerund is a noun used as a verb...................................

10. In the sentence in Question 8, the word "cat" is in the nominative case........

MULTIPLE-CHOICE QUESTIONS

11. A female fox is called a
    1. Foxe   2. Foxen   3. Vixen   4. Foxee   5. Fox

12. In the sentence, "The experiment showed the staff the cause of the trouble", the word
    'staff' is:
    1. The subject   4. A proper noun
    2. The direct object   5. In the nominative case
    3. The indirect object

13. In the sentence, "We used a Weston voltmeter, a most accurate instrument", we have an
    example of
    1. Apposition   4. The indirect object
    2. Double objectivity   5. The passive case
    3. The accusative case

14. Which of the following is a "source" type of possessive case?
    1. An hour's leeway   3. Leonardo's Last Supper   5. Carter's pills
    2. The cow's milk   4. His Father's house

15. In the expression, "Those were the good old days", the word "those" is
    1. An article   4. A demonstrative pronoun
    2. A relative pronoun   5. Demonstrative adjective
    3. A reflexive pronoun

16. In the expression, "They have only themselves to blame", we have an example of
    1. A reflexive pronoun   4. Disagreement in number
    2. A reciprocal pronoun   5. The possessive case
    3. A gerund

56

17. The sentence, "If one overloads a circuit you may have a blown fuse", contains the following error:
    1. Disagreement in number
    2. Collective noun used
    3. Misuse of conditional form
    4. Disagreement in person
    5. "One" and "you" are in wrong places, and should be exchanged _____

18. In the sentence, "We should all try to be more careful", the word "should"
    1. Should be replaced by "would"
    2. Is an example of an auxiliary verb
    3. Is a reflexive verb
    4. Is a copulative verb
    5. Is part of a compound verb "should try" _____

19. "Be I able, I shall gladly assist". This sentence
    1. Is in the subjunctive mode
    2. Is grammatically incorrect
    3. Should be rewritten "Were I able, etc."
    4. Should have "shall" changed to "will"
    5. Should be rewritten "If I were able, etc." _____

20. "His bungling caused a wrong reading to be registered by the recorder." In this sentence we see
    1. An example of the progressive form of a verb
    2. A gerund
    3. An example of apposition
    4. An example of the use of auxiliary verbs
    5. An example of a demonstrative adjective _____

OTHER QUESTIONS

21. Rewrite the following sentences (if you think it is necessary) in correct form.
    1. He went because he wanted to.
    2. George is going to definitely run for public office.
    3. Inasmuch as he wanted to go to the pool to swim, he went.

22. Complete the following expressions without using cliches.
    1. Soft as _____
    2. Dull as _____
    3. Weary as _____

23. (a) Rewrite the following sentence correctly. (b) Identify each part of the corrected sentence. (c) Identify the sentence type in its corrected form.

    The Professor, to the class, discussed some inconsistencies appearing in the speech.

LESSON TC-4
Writing Technical Reports

Introduction

The writing of reports is one of the more painful aspects of professional life for most engineers. To some otherwise competent men, the months of research, development, and design required for a new product are infinitely easier than the few hours or days needed to describe this work in a technical report. This abhorrence of writing is a curious phenomenon, for it is found in many men who have no hesitancy in *talking*, sometimes to the boredom of their fellows. It is shared by students of engineering, who generally put off term papers until too late, and one may suspect that the dislike of writing among engineers descends directly from similar repugnance while at school.

Report writing actually is not difficult at all for anyone who is literate in the English language. The trouble for most people is psychological, like stage-fright, and it can be overcome by a little insight and practice in the matter. It is the purpose of this lesson to indicate something of the nature of a technical report, to provide practice in the organization of typical reports, and, hopefully, to give rise to an attitude toward technical reports that will erase the dislike — that is, to eliminate the feeling that writing is a burden that no decent engineer should be asked to assume.

What is a Technical Report?

As we have mentioned several times already in these lessons, the sole product of the efforts of engineers consists of organized ideas. An engineer does not make anything, but he supplies the plans and information to others which enable them to make things.

For the purposes of this discussion, the work of the engineer may be divided into four phases. These are the following:

1. First, the engineer must form in his mind a clear idea of the end to be accomplished by the device or system he must create.

2. Next, he must review a number of physical phenomena which may be suitable for application. For example, if power must be transmitted from one point to another, he should decide among mechanical, hydraulic or electrical transmission. If he requires sensing the position of a machine part, he might have to choose among mechanical contact, optical detection, or some electrical effect.

3. Having decided on the physical phenomena to use, the engineer must make what is called the "reduction to practice." That is, he must design the actual nuts-and-bolts which will do the job required.

4. Finally, he must put his solution to the problem into a form which will guide the workmen who actually build the machine or system. The usual form is a drawing or drawings, which show shapes and give dimensions, tolerances, materials, and other needed information. Along with the drawings, however, verbal explanation, special notes, assembly sequences or instructions needed to build and adjust the device, may be required. Usually also a manual of instruction must be supplied

to the user of the device, and to technicians who service, adjust, and repair it.

In this lesson we are concerned with Phase 4, with emphasis directed away from the drawings and toward the verbal supplement. In later lessons we shall consider the several kinds of manuals, while our interest in the present lesson will be directed mainly toward written information developed during the course of design, intended to inform persons within the company of what is going on. In other words, in this lesson we are going to study the writing of reports.

Before studying report-writing in detail, however, the report itself might be put into clearer perspective by comparing it to other kinds of technical writing. In the first lesson of this group we noted that there are two kinds of prose writing: expository and narrative. Most technical writing is of the expository type. Its purpose is to explain and instruct, and only rarely is there the need to assume narrative style in order to carry the reader through a chronological experience.

You will remember that the field of technical writing can be subdivided according to the purpose of the material written. To review briefly, we have made the divisions as follows:

1. TECHNICAL REPORTS

Technical reports are descriptions of technical development work. Progress reports inform the reader of effort to date, and present status. Final reports provide a complete description of an entire project. The length or formality of a report depends on circumstances; it can be anything from a note on a memo pad, to a book printed in four colors on glossy paper to show the stockholders.

2. TECHNICAL MANUALS

Technical manuals are documents which accompany a piece of equipment to explain how it is operated, serviced, or repaired. They vary from the Owner's Manuals for automobiles, which tell little more than how to open to glove compartment, to multi-volume works that provide nighttime reading for submarine commanders.

3. TECHNICAL PROPOSALS

A technical proposal is a sales document. It describes how a problem can be solved or a device designed. It may be written by a designer for his supervisor within a company, or prepared in the name of the company for a prospective customer. In either case its object is to convince the reader that the project described is feasible and should be done. At the same time it should honestly note difficulties that may arise. The proposal writer must continue to live with his supervisor, and the company must still look for orders from its customers after the project is completed. If too many potential troubles are swept under the rug in a proposal, the writer (or the company) won't be trusted next time.

4. TECHNICAL ARTICLES

The professional reputation of anyone is helped if he gets into print. Companies like publicity for their products (as long as trade secrets are not let out) and reputation as a writer for a man can lead to job offers from other companies, and to raises to keep him at his home company.

The technical article must be "slanted" toward the backgrounds and interests of its intended readers. Examples abound in hundreds of publications, from newspapers and Popular-Electronics type magazines to esoteric scientific journals.

5. INSTRUCTION MANUALS

All technical writing is instructional in the sense that it gives information to the reader. In an instruction manual, the objective is usually broader than in the typical operation or service manual. The purpose is to provide knowledge or skill that will be useful in a definite area of work. Along with the instruction manual, writers may be called on to produce a syllabus (which states what is to be learned but does not attempt to teach it) or a textbook (which undertakes to carry the reader through the whole learning process, sometimes without the presence of a teacher.)

All of the foregoing types of production may employ illustrations. Photographs, drawings, diagrams, curves, and other graphical aids-to-understanding should be used freely whenever they will contribute to clarity and efficiency in the presentation. The writer is the logical person to coordinate the efforts of artists and draftsmen who do the graphical work, and he should know the capabilities and limitations of graphical aids. If the writer can also draw, he can help immeasurably by providing sketches to guide the illustrators. If the writer is also a skilled draftsman, the whole production can be given a consistency almost impossible to attain with separate draftsmen.

The eighth lesson of this group discusses the use of pictures in technical productions — not to make an artist out of you, but to make it easier for you to work with and direct the activities of artists.

The last lesson of this group discusses other functions closely related to writing. These are technical editing, and the preparation of films and other audio-visual media. The editor bears the same relation to the writer that the shop inspector does to the workman. Editing is no reflection on the writer, but is simply a stage in the manufacture of a written production. Knowledge that an editor will review his work gives the writer freedom to concentrate on the creative aspects of his work. The production of a technical film, whether for sales, morale, or training, must start with the writing of a manuscript — the *scenario* — which describes the production scene by scene and gives the narration. A good technical writer may be called on to prepare the narration script, as well as work with other personnel on the planning of the film.

Who Writes Technical Reports?

Technical writing is a profession in its own right, and in large companies full-time technical writers are employed to write proposals, manuals, and similar documents. Most technical *reports*, on the other hand, are written by the engineers and technicians whose work is described. If you are employed in a technical capacity, the most likely type of writing that may be assigned to you will be a report of what you are doing. A good job on this kind of report will obviously not only put you in a favorable light as a writer, but will sell you to your boss as a good engineer or technician. In the cold objective language of a report, you have a fine opportunity to put all your abilities on the record.

The reasons for asking engineers to write their own reports are practical. A report may describe the more complex details of a project, so that only the man who has developed them can fully and accurately describe the result. A professional writer with only a nominal background in the subject-field might require too much of his own and the engineer's time to gain the necessary understanding for writing a report. As we shall see later, the study of reports is one of the sources of information for writing manuals and other "second generation" documents.

There is also an advantage to the engineer in writing his own report. Putting thoughts into English on paper forces the writer to review and criticize his thinking, just as teaching clarifies the teacher's own ideas in the process of teaching them. In the course of writing a report, the engineer may discover errors in reasoning, inadequate data, or even conclusions difficult to defend.

Drawing a picture of something under design is a part of the design process, for it visualizes clearly what may have been imagined only vaguely. In a like manner, a written description of the many mental steps involved in design puts the whole process into better perspective.

In spite of the many advantages to the engineer in writing reports about his work, most technical men view the job with distaste, and try to get it done as quickly as possible in order to get back to their slide rules and drafting tables. It is a strange thing that the engineer who readily acquires skill with a computer objects to learning the more basic skill of report writing. Years ago when the writer of this lesson interviewed many chief engineers to ask their opinions as to what was most urgently needed in engineering curricula, the need for instruction in writing was always placed high on the list.

It is not easy to pin down the reasons why engineers don't like to write. Perhaps in high school their interest was oriented towards science and mathematics, and English teachers with little knowledge of science failed to gain their respect. In most technical colleges there is little emphasis placed on technical writing; even leading colleges with strong humanities departments usually emphasize literature and other "appreciative" subjects rather than writing. They tend to make English a "spectator sport" rather than one of participation.

Reasons put forward by some engineers to explain their poor writing ability include the complaint that they are not accurate in grammar, or cannot produce elegant prose like a popular novelist. The facts in the matter are quite the opposite. The use of "elegant" prose can hurt a technical report, and the man who uses very sloppy grammar probably isn't a good engineer or technician either.

The equipment needed to write technical reports includes a normal vocabulary, the ability to use understandable grammar which does not offend the average reader and, most important, the ability to think clearly and organize subject matter in a logical manner. Clear and logical thinking is — or certainly should be — one of the characteristics of an engineer. He must think in this way to do his job, and it is absurd to imagine that this ability suddenly deserts him when the time comes to describe what he has done. All he has to do as a report writer is to put on paper the steps he took in doing the job.

Tools Required

The tools used in the actual writing of technical reports are:

    1. Knowledge of grammar on the working level, and

    2. A basic style of prose writing.

You did a review of grammar in the preceding lesson. When in doubt you should refer to that lesson, or to any of many good general textbooks on English grammar. There is no more disgrace in looking up a grammatical form, than in getting a formula from an engineering handbook.

The *style* of prose writing is important because it gives life to the report. The most perfect grammar is worthless if the material is so dull that the reader is bored, or skims it while looking for some kind of conclusion.

Style in writing is difficult to teach, but it can be acquired by reading examples of good English prose. By this is not meant the sensitive writing of a great novelist, whose product is literary art, but the lucid explanatory style found in publications like *Scientific American* or *The Reader's Digest*. By reading well-written English, you can improve your own style. The process will probably be almost subconscious, and you definitely should not try to remember and copy specific phrases. Rather, you should do this reading as a kind of "sponge" which absorbs everything, and applies it to its own use after internal digestion.

Along with style, the report also needs good *organization*. We shall give examples of how reports are organized, but here we must again point out that organization is supposed to be a basic skill in engineering work itself. If your thinking is organized enough to carry out a project, then you need not worry about being able to organize a report about it.

Different Kinds of Technical Reports

A. PROGRESS REPORT

During the course of a design development or an investigation, management may want to know how the work is proceeding in comparison to the original time estimates. Although such a report usually is prepared for the person who originally authorized the work, the writer should not assume that his reader has the whole project firmly in mind, and can just take up where the last report left off. Business executives forget the details, and appreciate a report which brings them up to date on preceding reports, like the synopsis that prefaces each installment of a serial novel. The report should start by outlining the whole project, including the original assumptions and cost estimates. It should then detail the advance made since the last report, such as facts learned or design phases completed, and the funds expended in contrast to cost estimates for this stage. The report should include an estimate of the work remaining to be done, including any revisions in estimated time and cost.

B. FINAL REPORT

The final report is similar to the progress report except that it summarizes the total work of the project, presenting final conclusions, designs and costs. It is usually longer and more formal than a progress report, since it is likely to be read by more and higher-placed people in a company.

C. EVALUATION REPORT

An engineer is often called upon to make a study of an existing product or process, to assist a company considering its use or perhaps its discontinuance. Such a report, like the articles in consumers magazines, should make a fair appraisal of the subject, considering both good points and faults on comparison to similar items, and make a formal recommendation that it be accepted or rejected.

The writer of an evaluation report is usually not the responsibile decision maker, but when his report is the chief source of information for that person, his conclusions may have great effect. An engineer whose conclusions have been proven to be sound in a number of instances displays the kind of judgment that is constantly sought in all companies in its executives.

D. INVESTIGATION REPORT

An investigation report is somewhat like an evaluation, in that the writer collects pertinent facts and draws a conclusion. The investigation is usually of some specific occurrence, such as an accident or the failure of a machine or process. The writer should first describe in detail exactly what happened; second, undertake to find the reasons for the incident; third, describe the state of affairs afterwards and, sometimes, offer his own conclusions as to the basic causes and how similar accidents could be prevented in the future.

Some of the most painstaking investigations are those undertaken by the Federal Aviation Agency following major air disasters. The pieces of a shattered aircraft are collected and assembled on a hangar floor in their proper relationships, and the state of stress, temperature effects, etc., on each are studied. The report of such an investigation may run to hundreds of pages and require many months to prepare.

Other reports may be so short that they can be written on one sheet of paper, or perhaps made by filling in a printed form. Somewhere between these extremes will be found most reports of this kind.

The writer of an investigative report may find himself under pressure from persons involved in the incident, to slant it so as to reduce blame or embarassment. He should avoid any kind of whitewash, and should be fair to all to the best of his ability.

E. PROCESS REPORT

When design work results in a specific program for a process, a report on this must be prepared, usually as the basis for an instruction manual to be written later. Such a report is written as source information, and is usually a very simple narration of the steps required in carrying out the process.

A computer program may be regarded as a specialized, encapsulated process report for a calculation.

Who Reads Technical Reports?

Any piece of writing has for its object the creation of a specific impression in the mind of a reader. Clearly the means used for this purpose must be chosen with the reader in mind. Usually the reader of a technical report is a man with an engineering background who expects the report to be concise and well organized. For him, flowery prose is wasted; what this reader wants are the following:

1. A document which clearly and accurately provides the desired information.

2. A document suitable for reproduction to use in informing other people as to the situation. (This means that it should not be narrowly directed to one man, but should contain enough background so that others not as well acquainted with the subject can understand it.)

3. A document which can be understood when read years later, so that it can be filed for later reference. (This requirement is similar to #2 above.)

The person for whom the report is expressly prepared may be called the "primary" reader. He is usually a supervisor who understands the subject as well as the writer does, and merely needs to be filled in on the details to date. In order to qualify under points #2 and #3 above, however, the report should contain supporting discussion and facts, enabling the primary reader to pass it on to others in management who may not have his detailed technical background.

When it is known that a report will be read by persons with particular backgrounds and interests, the report may be "slanted" towards their viewpoints. For example, a report written for the engineering department might stress new design concepts; one written for the service engineer would stress accessability of parts and ease of replacement; one written for sales would be mainly concerned with ease of operation of a product and with the quality of the results.

For a new product, technical reports provide the information used by the marketing division of a company in preparing advertising, brochures, and operator's manuals. Good reports will help the technical writers who prepare these documents, resulting in overall benefit for the company, and in less engineering time taken up by interviews with the technical writers.

In making basic decisions about products, management must rely heavily on technical reports. A report which is clear and understandable will obviously have a greater influence on management than one which is poorly written so that it fails to give a fair presentation of its contents. Like all other written documents, technical reports have a "selling" function. When the "customer" is a company president or chief engineer, the "sale" can be very important.

Sources of Information for Technical Reports

The technical report is usually written by the man whose work has generated the information presented, so that there is no interviewing of others needed, except possibly that of assistants or technicians. During the course of his work, a competent engineer will accumulate a file of information for use in reports. This might include the following.

1. Copies of technical articles consulted prior to starting work, and notes and abstracts of other reports which influenced the project.

2. Notebooks containing day-by-day notes in detail on the progress of the project. (Such notebooks also are valuable in establishing patent priority.)

3. Directives and memoranda relating to the project.

4. Drawings, sketches, graphs, blue-line prints, and other pictorial material developed in the course of the project.

5. His reasoning for various decisions made, and general ideas concerning the project.

The total volume of material contained under the above categories may be quite large. The report writer's job is to distill from this the important facts and conclusions. Details, such as graphs of a particular test, should be given only when the writer wishes to illustrate by example a test procedure which may be important in its effect on the answers. For example, if points lie on a curve with little scatter, the curve is more reliable and this fact might be important to the reader.

Organization of Material

The first problem facing the report writer, once he has assembled on his desk all the documents and pictures bearing on the subject, is that of organization. The question is, "What is the best way to build up in the mind of the reader the same overall understanding of the project that the writer has?" One approach to an answer might be to present the project in the same way that the writer himself experienced it, that is, chronologically. Many reports are written in this way, almost as a story. Although this method makes it easy on the writer, it may become tedious for the reader, especially if the project went into several blind alleys before final success was attained.

A second approach in writing is to *subdivide* the report according to the *major divisions* of the project, and then treat each division either chronologically, or by the "concept" method which will be described next.

The third or concept approach to a report is to describe the development in terms of the reasoning that went on during the development. The initial concept might generate several ideas which had to be analyzed until the best was determined. When a project has run into several blind alleys, the concept method of writing has a great advantage in that it provides a record of thinking which may serve to prevent a repetition of the same errors later.

Let's illustrate the foregoing three approaches by a concrete example. The project concerns the design development of an electronic facsimile system for use in sending photographs over the telephone, without any electrical connection.

The engineer assigned to development might use the *first approach* and write a chronological report according to the following outline.

I. Initial work:
   1. Preliminary design planning.
   2. Assembling the design group.
   3. Assignment of space and equipment.

II. First month:
   1. Initial experiments on resolution using 5″ CRT as a scanner.
   2. Construction of breadboard circuit.
   3. Tests with microphone vs magnetic telephone pickup.

III Second month:
  1. Design conference. Decision to use 9″ CRT to obtain greater resolution.
  2. Design of low pass filters.
  3. Resolution testing.

IV Third month:
  1. Final circuit configuration.
  2. Layout of printed circuit-boards.
  3. Internal geometry of components.
  4. Outside appearance design.

V. Fourth month:
  1. Final tests.
  2. Summary evaluation of performance.

If the writer chooses to follow the *second approach* ("subject" approach) in his organization of the report, his outline might be as follows.

I. General summary of the problem as presented to the design group.

II. The transmission system:
  1. Selection of flying-spot scanning tube.
  2. Selection and placement of photomultiplier tubes. Reflection problems.
  3. The transmission circuit. Transmission loudspeaker.

III. The reception circuit:
  1. Microphone vs inductive pickup.
  2. Reception circuitry.
  3. Problems of biasing and brightness.

IV. The sweep circuits:
  1. Horizontal sweep circuit.
  2. Vertical sweep circuit.

V. Physical arrangement of components and external design.

The writer electing to use the *third approach*, the "concept" method of writing, would probably like to start with a little preliminary history prior to the actual start of design work. His outline might be as follows.

I. The economic need for the device. Potential uses, and users. Need for self-power and portability. Conditions of use. Sophistication of the users. (These considerations must be kept in mind by the designers throughout their work.)

II. Reasons for selecting 5″ CRT.

III. Design of transmitting amplifier. Why use frequency modulation.

IV. Design of receiving circuits. Trials of microphone and inductive pickup.

V. Resolution tests. Decision to change to 9″ CRT.

VI. Problems relating to the deflection circuits.

VII. Problems relating to the power supply.

VIII. Preliminary tests on color transmission (for benefit of later project).

IX. Design considerations leading to internal layout and external appearance of unit.

Which kind of organization is best? It depends on the nature of the project, and which aspects the writer wants to emphasize. The chronological method presents events as they occurred, with equal emphasis. If the report concerns a series of tests extending over considerable time, the chronological method might give the truest picture. For example, in chemical or biological research, significant gradual changes are best noted in this way. The effects of diet on rats, or of various temperatures on beer fermentation, might be clearest when presented as a story.

The "subject" approach enables the writer (and reader) to concentrate on one thing at a time. There is danger that the reader may lose sight of the overall problem in his attention to detail — i.e., miss the forest for the trees. The writer must guard against such an accident by writing introductory or bridging paragraphs between the parts to show the relationships. The use of block diagrams, with each block represented by a section of the report, is helpful.

It might be noted that the design of the curriculum in a college essentially employs the subject approach, in that students concentrate on various phases of their education, one at a time. Such curricula have the danger just noted — in this case, that the student may fail to get a clear integrated picture of his whole education.

The concept approach may be described as the most sophisticated of the three and, in general, the one requiring the most thought and organization by the writer. A good report arranged in this way has much in common with a good textbook, in the way that it leads the reader through stages of reasoning in order to build up his mental concepts. In chronological and subject styles of writing, the reader must usually form his own conclusions. In the concept style, the reader is led to the conclusions the writer desires, by the way the report is organized.

A chronological report might be described as a Reader's Digest kind of condensation of a diary of the project. A subject report is usually a series of chronological reports concerning different parts of the project. A concept report gives an integrated picture of the whole project, with emphasis on the reasons; it gives the "why" as well as the "how".

In most reports it is desirable to include an *abstract* — a short paragraph which summarizes the report and presents its main conclusion. The abstract of necessity is in "concept" form. Abstracts serve several uses. When many reports cross an executive's desk, he can read all the abstracts and then decide which ones require his more detailed attention. Abstracts can assist an engineer at the start of a project in getting an idea of how much work has already been done. When a report has been filed, the abstract can serve to remind a man who read it previously as to its contents.

In some companies, particularly the very large manufacturing firms, it is desirable that all reports adhere to a definite order of presentation. Such a policy creates consistency when there are many report writers of varying inclinations, and also helps the inexperienced. The following guide is similar to that published by many companies.

1. Title page:
The title page contains necessary identification: name of report, report number, name of writer and his department, and name of company. In reports prepared for a Government agency as a customer, there are usually other pieces of information

required, but in these cases the writer will have an elaborate printed form for his guidance.

2. Letter of transmittal:

This is a formality which is gradually disappearing from reports. It is addressed to the principal reader of the report and says something like this: "The final report on Project 23-7, High Altitude Missile Detector, is herewith submitted. Very truly yours, ..."

3. Summary:

The summary or abstract is a synopsis of the main points of the report, with emphasis on its conclusions or recommendations. Highly technical terms are avoided, so that it can be read by a layman as well as by one who is technically trained. Many people may read the summary only, and attempt to judge the validity of the conclusions on whatever technical data is given in it. In the case of reports in which the numerical data is highly important, careful samples of this should be included in the summary, so that even the quick reader will have a chance to make a fair judgment and not be tempted to reject the conclusions on a basis of insufficient supporting evidence.

The summary has a dual function — that of preparing the way for the reader who intends to go thru the whole report, and that of serving as a complete encapsulated report for the very busy reader.

4. Table of Contents:

The table of contents lists the divisions of the report in page order. If an alphabetical index is to be included, the table of contents can include only the major divisions; if there is no index, the contents should list most of the sub-sections of the report so that a reader can find any given topic by scanning through it.

5. List of Illustrations:

If the illustrations form an important part of the report or if there are many illustrations, they should be listed. If there are only a few, this page may be omitted. The criterion is whether there is needed convenience for the reader in having such a list.

6. Introduction, or Preface:

The preface to a report serves the same function as that written for a book. It should introduce the report by describing the background of the project, and the circumstances under which it was started. A preface is also the proper place to make acknowledgments for help given by people outside of the immediate project staff. In a book, such acknowledgment is an appreciated courtesy. In a report going to the executives of a company, the mention of a name in the preface can be of considerable practical value to the career of the person mentioned. Care in acknowledging help, even in a small amount, will be appreciated and will create more cooperation next time. The preface should not overlap in its contents with the material presented in the summary.

7. Main Body of Report:

This section constitutes the greatest part of the report, and may be subdivided into sections, chapters, or in any way suitable.

8. Conclusions:

An essential feature of teaching is repetition. The conclusion of a report has the function of driving home the important ideas developed in the main body by stating them again with proper emphasis. The conclusion gives a view of the "forest" in contrast to the detailed discussion which may emphasize the "trees".

The information presented in a "conclusion" may be the same as that given in the summary. The "conclusion", however, assumes that the reader has just finished the main report, whereas the summary is written for a reader "starting cold" who must first have a condensed statement of the problem and of the method used to reach the conclusion.

9. Bibliography:

A bibliography is a list of other books, reports, articles, etc., consulted in the course of the project, or in writing the report. The report may contain a quotation or data from such a source, or it may paraphrase information written in other words. Each reference should be identified so that it can be found readily in case the reader wants to study the material referenced in more detail. Material listed in a bibliography need not be restricted to publications. Unpublished memoranda, notebooks, or private letters can be referenced if they are filed so as to be available to the report reader.

10. Appendices:

Frequently in the course of a narrative discussion, it is undesirable to interject supporting data or ideas which might interrupt the flow of thought. If the auxiliary material is short, it can be put in the form of a footnote. If such material is lengthy, it is better to place it in an appendix at the end of the report. For example, the derivation of an equation used in the report could be appropriately put in the appendix, if the equation itself is not commonly known or if the derivation includes assumptions or approximations that should be clearly understood.

Tables and other statistical data which support figures presented in the report should be included. In general, material should be in the bibliography if it is readily available, and in an appendix if the reader cannot obtain it elsewhere. Data generated in the reported project should be given in an appendix, or perhaps referenced to the project files, as is deemed best.

In general, material which might interfere with the continuity of reading a report but which is still needed after the basic ideas have been made clear to the reader, should be placed in appendices.

11. Index:

The index is an alphabetical list of all topics and important items mentioned in the report. In deciding what words should go into an index, the writer must consider what the report reader might like to know about. Conclusions, methods, apparatus, and materials are typical of things to go into an index. The index is particularly important for a report written in the "concept" style, since there may be no logical chronological way for a reader to search out a particular matter.

In some long reports, as in books, there may be two indices — a subject index, and a name index, the latter listing names of people whose work is referenced, or who are of importance in other ways to the project.

Style in Report Writing

Your first reaction to the title of this section of the lesson may be, "Why worry about style?" A Technical Report isn't a novel or a piece of poetry. It's a practical document designed to get information across, and it doesn't need any frills.

Such a reaction is right to a degree. A report should be written to efficiently convey information and impressions to busy readers, and literary elegance and padding can be annoying to them. On the other hand, a report, as a means to an end, should be regarded as a project in itself; its goal is to create a desired mental state in the reader. Words and sentences are the tools which form thoughts in the readers mind. If they are used well, the desired mental state is created and the report is successful. The way words and sentences are used constitutes the *style* of a report.

What is style, and what kinds of style are there?

Many scholarly books have been written to answer these questions. For our purposes, we can explain a number of ways of writing which may be described in general by the word, *style*.

Any writing can employ the past tense or the present tense. In a report, a description of experiments, discussions, and decisions which occurred during the course of work should properly be given in past tense. The conclusions, having current validity, may then be presented in present tense. There's a television news reporter in Los Angeles who likes to describe events in the present tense. For example, in describing a robbery that occurred the night before, he might say, "At 10 PM the suspect enters the store, where he is seen by Mrs. Smith. She calls the police, and in ten minutes the building is surrounded." The news reporter's object is doubtless to give an air of immediacy to his reports, but the overall effect is often confusing. But that the news reporter does this is a matter of *style*.

As in any narrative writing, the *conditional tense* should be used to distinguish between what is established, and what *may* be, depending on circumstances. For example, "An arc would form whenever the voltage exceeded 7.3 kv."

An important style decision concerns the use of formal third person writing, or the use of the first person. Most reports are formal. The writer and other project members are spoken of as "the staff", the "chief investigator", "the audio technician", etc. For example, a report would read, "The high voltage was turned on", and not, "We turned the high voltage on". In formal third person, the reader is never addressed directly as "you", but always in the third person as "the reader". E.g., "The reader will readily see that..."

In many textbooks, the reader is smoothly led through a discussion by using "we," which suggests a teacher accompanying the student through an intellectual maze. "We first connect the airhose. Then, after we are sure the pressure is right, we carefully open the main valve..." This kind of style is valuable in instruction manuals, and we shall refer to it in detail in a later lesson.

In textbooks of the more formal kind, "we" is not used. However, in these Grantham lessons, we not only employ it almost constantly but we often take a further step into informality, by addressing he reader directly as "you". Our object in using this style, of course, is to try to create a personal classroom atmosphere, one of the main objects of a correspondence lesson.

Technical reports can use "we", especially in small companies where there is a personal relationship between writers and readers.

The style of a technical report must be chosen with the reader in mind, just as the technical content must be matched to his interests and background. If the reader is a highly educated man, the report can use an expanded vocabulary and sophisticated sentence construction. If it is addressed to practical men of limited education, it must be in a form that is familiar and comfortable for them. Here, as always in writing, the fundamental criterion is that the document do the job it is designed for. As a report writer, you should keep in mind the fact that your reader also may have to prepare his own report for his superiors. His source of information will be your report. If he can also use your organization and lift your words and sentences bodily, his work will be easier and his opinion of you as a writer will be higher. However, if your boss pays you the compliment of using some of your report directly, don't go around boasting to others of this fact; word will get back to him in an embarrassing fashion, and you may be in the doghouse! There have been instances when a subordinate writer has been asked to prepare his superior's report, thereby entering into the very honorable profession of ghost writer. So, if you know that your report is going on upstairs, modify your style by thinking of who will receive that *next* report. By treating your boss as he treats his boss, you will be paying him a subtle compliment that will be appreciated.

Practice Exercises

We learn how to write by writing, not by reading about it. So far in this lesson you have been doing a lot of reading; now the time has come to write. So, following the advice of a famous fiction writer, you should take the following simple steps: 1. Pick up a piece of paper and place it before you (or put it in your typewriter). 2. Pick up a pencil (or position your hands above the typewriter keyboard). 3. Start to write.

Before turning to the practice exercises, however, some comments on the mechanics of writing are in order. We mentioned pencils and typewriters. If you can use the typewriter, you have a great advantage over the scribbler. It is faster than longhand, and the results are more legible — both for yourself to reread for corrections and changes, and for others including the stenographer who will prepare the final draft. Much of the dislike of writing is really dislike of the painfully slow pace of longhand composition. The writer of this lesson was fortunate in having a typewriter to play with at the age of seven, and, though still a two-finger typist, he can write about as fast as his mental processes will permit. When he occasionally is forced to write longhand, the result is painful and frustrating — the end of a sentence is forgotten while the first part is being written. So if you cannot type, you are most strongly advised to learn to use the typewriter without delay. Take a night course in typing if you can, but in any case, get a typewriter and start using it — if in no other way, by means of the hunt and peck system. Within a week you will be going as fast as longhand, and in a month you may never want to write in longhand again.

There are some "writers" who do their work by means of a dictating machine. Many find this method entirely successful. The detective story writer, Earle Stanley Gardner, for example, dictated his Perry Mason stories, and reportedly did not even read the manuscripts before they were sent to the publisher. The writer of this lesson has tried dictating but found it annoying to be unable to pause and reread the last few words before going on. Such things, however, are matters of personal preference. Each writer should select the method of getting his ideas on paper that is easiest and fastest for him — whether by pencil, typewriter, tape recorder, or using hammer and chisel on a stone tablet.

Let us now return to the subject of practice exercises. First we will consider a report on a mishap in the shop. Your boss wants to know if it was due to some malfunction of equipment which should be corrected, or whether it was just the result of clumsiness on the part of the workman involved. It is your job to interview the workman, and any witnesses to the incident, and put the facts down on paper so that the reader can form his own conclusions. You may come to a judgment yourself in the course of your investigation, and this judgment may properly be made a part of the report, although you should be careful to identify it as such and not allow it to color your presentation of the evidence.

In approaching an assignment of this kind, your first problem may be to get the cooperation of witnesses. The man directly involved may fear that he'll be fired or disciplined for his mistake, and so he may try to throw all the blame on someone else, or on the equipment involved. A witness may bias his account, depending on whether he is a friend or rival of the man directly involved. You will find often, in your role of investigator, that you are also acting like a detective in seeking clues, and like a psychiatrist, in getting people to talk freely.

For each practice exercise, refer to our comments under "Discussions of the Practice Exercises," which is the section of this lesson following the exercise themselves.

PRACTICE EXERCISE 1:

First, let's get to the interview. Your boss called you into his office at 8:25 AM and the following conversation took place.

BOSS: We've been having some trouble in the circuit board department. At first it looked like a malfunction in the solder machine, but Stevens swears it's working right. I intended to look into it myself, but I've got to go to Frisco for today and tomorrow, so I wish you'd look into it and write me a report for Thursday.

YOU: I heard that there's been problems with rejects. What's the story? (Your first source of information is the man who gives you the assignment. He knows something about the matter, or he wouldn't want you to write a report.)

BOSS: Well, as near as I can get it, it's where the boards go over the solder fountain. Sometimes everything doesn't get tinned. Warren had it checked, and ordered the belt slowed. Then last night inspection began reporting board rejects — too much heat for the transistors. Well, you look into it and see what you can find out. I gotta get the airport bus.

Act II of our little drama is laid in the solder room. At center stage is the soldering machine, where circuit boards with components mounted in place are passed over a tank of liquid solder. At the center a pump sends up a fountain of fresh solder, free of oxide scum which forms on the surface. The fountain is really a kind of mound or wave, rising an inch or so higher than the general surface. As the circuit boards pass over, the wave contacts the undersides and coats the etched copper and the wire-pigtails from resistors and transistors which stick through from the other side, soldering them together. Or at least, that's what is supposed to happen. The characters are you, George Stevens the foreman, and Billy White who runs the machine. George is uneasy, and regards you as an outsider sent in from the front office to harass him. Billy was late this morning, and hopes that if he is extra cooperative, his crime will be forgiven.

GEORGE: Well, I did as Warren said, and that's the story. I hope you can find out what's wrong; it isn't doing production any good as it is now.

BILLY: Anything I can do, you just name it.

YOU: Billy, what do you think caused the overheating?

BILLY: That belt just goes too slow. I told Mr. Warren at the time — you slow down that belt, you're gonna cook them transistors.

YOU: But then why didn't it solder when it was going faster?

BILLY: Beats me, boss. Never had no trouble like this before. Mebbe somepin different with the solder.

YOU: When did the malfunction start?

BILLY: What's that, boss?

YOU: When did you first notice that it wasn't soldering everything?

BILLY: Oh, last Thursday or Friday. I told Mr. Warren, didn't I George? Told him right off, there's something wrong he better fix.

YOU: And it was still giving trouble Monday?

BILLY: Yessir, but I told him and told him. What else could I do?

YOU: You did quite right, Billy. Have you got one of those boards that didn't get all soldered?

BILLY: Well — here, lemme see. (Searches on the bench, finally finds a circuitboard under some greasy rags.) 'Spection come and took 'em yesterday, but here's one.

YOU: (Looking over board) It's kinda greasy, isn't it Billy?

BILLY: (Defensively) Well, I was changing the belts down under the drive. Musta got some oil on my hands and wiped it off.

YOU: Was this after Warren ordered a slower speed?

BILLY: Well, . . . yes, that was after.

YOU: Did you do any working on the drive last week?

BILLY: Oh no, never touched a thing.

YOU: I mean, like oiling the bearings.

BILLY: Oh, sure. I keep them bearings oiled right.

YOU: And you always wipe your hands off on the rags before handling a board?

BILLY: Always. No sir, ain't no dirt or grease around here. Right, George?

GEORGE: (He's getting the drift now.) That's one thing you gotta be careful of, Billy. Any grease on those boards, and the solder won't stick at all.

BILLY: That's right, George. (He now knows what happened too.) I keep things real clean. Them old rags, just there by accident. (Hurries to remove rags.)

YOU: Well, thanks a lot Billy. I think I get the picture.

BILLY: Nothin' I did wrong, boss?

YOU: Not a thing. Just keep things clean, like George said. I think you can clear it up. (You and George walk away.) George, it isn't my job to make suggestions, but I think we both know what happened.

GEORGE: Don't be too hard on Billy, he's a good man. Just a little absentminded. I think one of the kids was sick last week.

YOU: Sure, we all have our moments.

GEORGE: I think Warren will have the speed brought back up to normal. Thanks a lot for helping us out.

YOU: Oh, you would have found it. Just luck, me with my dumb questions. Well, I'll get back and type up a memo. I'll send you a copy. (Exit)

Now that you have investigated the situation, write that report in 100 words or less for your boss, with copies to Warren and Stevens.

PRACTICE EXERCISE 2:

For a second practice exercise, you will be asked to write a short progress report on an infra-red talking light beam. The report will discuss only one portion of a project, and is concerned with an experiment on the use of a flashlight bulb as a transmitter. First, let us present the *information* required in the report. Read the following to obtain the necessary details.

The transmitter of a talking light beam is a device which emits variable infrared radiation in accord with the output of an audio amplifier. There are a number of ways to produce modulated infrared. A light valve driven by a voice coil will modulate an infrared source as well as visible light, and this method is the principal one employed in the project. A week ago, however, the suggestion was made to Steve White, the chief engineer, that the light produced by a filament type lamp fluctuates when a variable current is passed through it. For example, the light produced by ordinary lamps has a 120 hertz "flicker". This is not apparent to the eye because of persistence of vision, but a photoelectric cell will detect it readily. The change in light output results from the fact that the filament temperature rises and falls with changing current. The amount of change in temperature depends on the fluctuation of current and also on the amount of stored heat in the filament. This latter effect is often called "thermal inertia". A very thin filament which has little stored heat will vary in temperature more than a thick filament will.

The suggestion made was that the filament in a small flashlight bulb, having a very small thermal inertia, might produce a usable amount of "flicker" or "ripple" into the range of voice frequencies — e.g., to 3,000 hertz. At normal operating currents, this ripple would constitute modulated visible light. At reduced current, however, the filament would become invisible, but would still produce modulated infrared.

Three days ago the Chief Engineer described the background to you, and asked that you make a test setup and see if the idea was practical. You made a simple setup, in which an audio amplifier output was connected to a flashlight bulb, and a photoelectric cell was placed nearby to observe the output. On the first test, which was made with sinewave signals, you found that he photocell registered twice the frequency that was applied to the lamp. The reason for this, of course, was that the light output became bright for both the positive and negative excursions of voltage applied. It was therefore necessary to supply a direct current on which the ac was superposed, so that there would be one cycle of brightness change for each cycle of applied signal. Once this bias current had been supplied, and a few bulbs burned out by applying too much signal during initial experimentation, the test was quite simple. Frequencies between 100 and 3,000 Hz were applied and the photocell output noted. The results were displayed in a curve of photocell output against frequency. For the first lamp tested, the output was unusable at frequencies above 500 Hz. Another lamp having a smaller rating was tried, and this was usable up to 1200 Hz. The very tiny lamps of the kind called "grain of wheat" from their size were next used, and this time the output was good up to the required 3 kilohertz. Unfortunately the total energy emitted was so small that the range of a signaling device using this lamp was only a few feet, even when the best optics were used. Your conclusion was that the idea was interesting and feasable in principle, but not practical. You offered a suggestion for increasing the light output, using the smallest lamps.

Now that you have the necessary information, write a report in about 100 words to the Chief Engineer, showing your experimental procedure and the results. The latter, of course, would be shown in curves for the various lamps tested which would be a part of the report. See if you can deduce a method for increasing the infrared output, and present this in your report as a recommendation for further experimental work.

PRACTICE EXERCISE 3:

This exercise is concerned with a type of report frequently asked for in large companies — a personal evaluation of a co-worker, for the guidance of the personnel division. Such a report may concern a man working under your direction, or it may be about your own boss. Such reports may be quite difficult to prepare because of personality factors. You may like the person and want him to be advanced regardless of his real value to the company, or on the other hand you may dislike him because he has treated you unfairly, and therefore be tempted to take the opportunity to get some well-merited revenge. In either case you should proceed with caution, because experienced personnel people are aware of all the tricks in the trade, and an improper report on a sensitive subject like this can become a bad mark on your own reputation. On the other hand, a bland reply that says nothing is also poor, since it brands you as a spineless Mr. Milquetoast.

Let's suppose that you are asked to evaluate your immediate superior, Lee Casperson. Casperson, a young PhD, is a brilliant engineer in the field of laser physics, and has headed up a design group of 23 engineers and technicians. He is 24 years old, which is a little under the average for the group, and ten years under your age. There has been a certain amount of jealousy on the part of older men who feel that their seniority entitles them to the promotion which management is considering for Casperson. You have been annoyed on two occasions by snafu's in which Casperson has displayed an ineptness in management, out of keeping with his technical ability. When you first heard that he was under consideration for the job of Technical Manager, a purely administrative position, your first reaction was one of relief. You felt that anybody would be an improvement. You also felt that he would not do well in this non-technical capacity and that his real genius would be wasted. Then, just to complicate matters, Miss Robbins, the vice president's secretary, whom you have occasionally dated, let slip the fact that if Casperson were promoted, you would be in line for his job. You believe yourself to be experienced in management and able to get on with the others in your group, so on a personal basis, you have everything to gain by kicking your boss upstairs.

On the basis of the facts given above, write an evaluation report on Dr. Lee Casperson, in from 150 to 300 words.

PRACTICE EXERCISE 4 (Editing or Rewriting):

For the practice exercises which follow, we are going to present some examples of mangled English, and examples of bad report-writing style. These will contain errors in spelling, grammar, and sentence construction, as well as unclear writing and the use of cliches. Your job is to bring order out of chaos, taking the part of a technical editor. We shall discuss Technical Editing itself in detail in a later lesson of this Series, but these examples are presented here because every technical writer must also be his own editor as he writes. So let's see what kind of literary repairman you can be.

Here is the opening summary placed on the first page of a detailed report on the design development of a multi-purpose testmeter having a digital display. Do a literary repair job on this paragraph:

The meter is built to measure voltage in six ranges and ohms in four ranges. It utilizes a highspeed counter which counts pulses in the megahertz frequency. The count is displayed by three nixie lamps on a digetal display, and the state of the counter is used to feed current increments to a junction (digital-to-analog conversion) where they are compared to the unknown signal. When a balence is reached, the oscillator is stoped so that the final count, representing the unknown volts, are displayed. This process is repeated ten times a second."

A *good summary* should highlight what the device does, and indicate in a very general way how it works. Also, correct spelling and clear sentences should be used. How many spelling errors did you see in the previous paragraph? The following re-write is quite an improvement though still imperfect in wording.

The DVM consists of an oscillator and counter, which converts pulses into digits. By doing it ten times a second, the readings are quick as a wink. In the digital to analog converter, the most significant place is represented by ten milliamps, the next significant place by one milliamp, and the least significant place by 0.1 milliamp, except that *binary* notation is used in the counter. When a frequency of one megahertz is applied, a count of 1000 is reached in one millisecond. The counter is set back to zero after 0.1 seconds, and the counting process begins again automatically. In this way the lamps indicate the unknown voltage for 99% of the time so that it appears as a continuous indication.

The following is another example of poor report writing. This time we are taking a look at part of a report on *life tests* made on an electro-mechanical relay. The fragment shown here is the description of the set-up and of the conditions under which the testing is done. This portion of the report might be written under the subheading, "Test Set-Up". Rewrite the following three paragraphs in a form you consider more appropriate.

"The relays to be tested are put in a hay-wire rig that holds them where the salt spray and other goodies can reach them easy. A cam on a geared down motor closes twice per second, which closes the relay and makes a counter work. This makes 173,800 cycles per day, or 1,730,000 cycles in the ten day test, which is enough for any relay.

"While one relay was getting this test, another one was put in a high humidity chamber and given the same treatment. Relay number three was doused with salt-water spray for one hour and then tested the same way, and relay number four was tested while being splattered continuously with saltwater spray.

"Whenever a relay conked out, the counter which is connected to one of the relay contacts naturally stopped counting, even though pulses were still being applied to the relay coil. In this way you could tell how long the relay lasted even if it went out in the middle of the night."

Discussions of the Practice Exercises

Practice Exercise 1: The interview brought out very clearly the cause of the trouble. The initial problem in making the solder stick was due to carelessness on Billy's part in handling the circuit boards with greasy hands. The grease smears prevented proper adhesion of the solder when the board passed over the fountain. The real blame, however, should be laid at the door of Stevens, the foreman, and to Warren who presumably is a department head. Stevens in particular should have found the original cause of the trouble, and not allowed his superior to order the speed change (or even to have recommended this, although the exact situation is not brought out in the interview). Stevens is understandably worried about his failure, and one may be sure that Warren will be too. The interviewer was wise in being modest about his part in uncovering all this incompetence. Blame is not for him to assign. He should, however, state the facts as he found them, and he should discreetly let his boss know that it was he who uncovered the cause of the trouble. Such a disclosure will obviously not improve the positions of any of the other three, but the report writer's *first* concern should be for the general good of the company, and his *second* concern should be for his own career. Protecting incompetence will help neither.

The report should first state the facts of the matter, and then indicate a recommended solution. No blame should be assigned, but you may be sure that the boss will draw his own conclusions. There are many possible ways or wording such a report; the following is one way:

To: A. R. Becerra

Pursuant to conversation of 3rd March, a discussion was held with George Stevens, Foreman Dept 27, and Billy White of Dept 27, on the subject of production troubles in the soldering of circuitboard assemblies.

On Thursday or Friday of last week, it was observed that solder failed to adhere to portions of the circuitboard. In the belief that the lack of adhesion was due to insufficient heating time, Mr. Kenneth Warren, Manager of Dept 27, ordered a decrease in the speed of circuitboards passing over the solder fountain. This change was made by Billy White. On Monday of this week, inspection rejected a number of circuitboards for faulty transistors. This fault was attributed by Mr. Stevens and Mr. White to overheating of the transistors as a result of the decrease in speed. Mr. White suggested that the initial problem might have been caused by a change in the composition of the solder used in the machine.

The writer examined one of the original faulty circuitboards, in which solder had failed to adhere. He observed spots of grease on the board at the points where adhesion had not occurred. Mr. White stated that his fingers may have become contaminated by grease in the course of maintenance work on the machine. Mr. White and Mr. Stevens are now both aware of the necessity for keeping the circuit-boards free of grease, Mr. Stevens indicated that he would discuss the situation with Mr. Warren, and that he was sure that the whole matter will be resolved.

Copies to George Stevens, Kenneth Warren.

In our example above of how this report might be written, notice that the wording is careful to avoid direct criticism, or to refer boastfully to the fact that the writer uncovered a simple cause which should have been found by the responsible people of Dept 27. Note that the report does not say that Mr. White's hands were greasy, but quotes him in his admission that this might have been the case. The report makes it plain that the situation is under control. If the Boss wants to call in Stevens or Warren for further action, he can do so at his option. (A wise boss probably would not do so, as both men have been sufficiently jolted already.)

This report will be filed, and a good mark made on the writer's record, both as a troubleshooter and a diplomat.

Practice Exercise 2: In this report the main information concerns the results. The text setup is so simple as to require no particular detail in its description; the Chief Engineer assumes that you can figure out such a test, or you wouldn't have the job of test engineer.

If you were writing the report in chronological style, you would describe your initial try without the biasing current. There are several reasons for not including this information. It has no bearing on the matter to be determined, and it is so elementary that it is slightly embarrassing to confess that you didn't think of it beforehand. You don't need to deny that you didn't provide bias current in the first place, but you can simply mention that this was supplied in the tests. If the Chief finds out from some gossipy technician, he will not be seriously concerned, since he probably can remember many slips like this in his own career. So here is one way of writing this report:

To: Steven White, Chief Engineer.

From: You.

Subject: Feasability study of lamp as infrared modulator.

Pursuant to conversation of 3 May, a test setup was made to investigate the infrared ripple output of various filament lamps over the audio frequency range. Fig. 1 shows the test arrangement. A is the lamp under test; B is a variable audio oscillator; C is a power amplifier; D is a series battery to provide direct bias current to insure unidirectional current through the lamp; E is a multiplier phototube; F is an audio amplifier; and G is an audio-level meter.

In each test, amplifier gain was adjusted so that the lamp filament was barely visible as a deep red at maximum signal. Frequencies in the range 100 Hz to 3,000 Hz were applied at constant a.c. voltage across the lamp, and the resulting signals from the phototube amplifier were recorded.

The results of the tests are shown in curves 1 thru 6.

It is clear that only the grain-of-wheat lamp produces a usable output ripple at the higher audio frequencies applied. With optical components available, it is estimated that a signal from this lamp could be detected at a distance of approximately twenty feet, which is much less than the distance requirement for the equipment under development.

It is accordingly concluded that currently available filament lamps are not suitable as infrared modulators in the range of 100 thru 3,000 Hz.

If a number of grain-of-wheat lamps are operated together, however, the infrared output and hence range will be correspondingly increased. For example, an assembly of 100 such lamps arranged in a compact 10 x 10 array, will produce a usable beam to a range to between 200 and 400 feet. It is suggested that such an array be constructed and tested for evaluation in comparison to other modulating devices under consideration.

Practice Exercise 3: Now here is a report that might be written in the situation described in Practice Exercise 3.

To: Engineering Personnel.

Subject: Confidential evaluation report.

From my observation as a member of the design group under Dr. Casperson, I would characterize him as a designer of great creative ability, qualified to originate technical projects and carry them through to successful completion. I believe that his genius in the technical field is so great that a change in assignment to administrative work would deprive the company of ability very difficult to replace. Personalities aside, I would believe that Dr. Casperson would be both happier and more productive in an assignment in which technical ability rather than managerial skill is called for.

Notice that the report given in the above example is couched in very complimentary terms, but it enables the personnel division to read between the lines, that in writer's opinion the candidate is not suited to management. The expression, "personalities aside", hints delicately that there may have been personal conflicts, and that you are trying to make your appraisal without letting them color your views.

Frequently, personnel evaluation reports are written very bluntly, especially when it is known that the personnel manager or other supervisor involved likes to speak plainly. When a report is destined for such a person, a carefully worded report like that here presented would make a poor impression, and the addressee might throw it out altogether as mere pussyfooting. For an extreme example of blunt writing, the following might serve:

___

Dr. Casperson is a very good technical man, but he fails as a manager. He has trouble communicating with the men under him, and he is a poor planner who will often overload some men while others are left without adequate guidance. In my opinion it would be a mistake to put him into managerial work. He is doing a good job where he is, but I would suggest that he be given some assistance in the planning and organization of work, or perhaps given a consulting assignment or some other work which will utilize his fine technical ability.

___

Before writing a report of this kind, you should be very sure that it will not get back to the man concerned. Very few people can stand criticism of this kind, and you might find that you are working under a very pained and ruffled supervisor. Obviously, good security for the writer is essential before valid personal evaluation reports can be expected.

An interesting practice in some companies is to ask the man himself to write an evaluation. This sort of report lays out a fine tightrope for one to walk, in balancing between self-praise and leaning over backward with criticism that just might be taken literally!

Practical Exercise 4: Turning our attention to the second report shown in Exercise 4, we suggest that this report represents a low ebb in report writing. It is uneven in style, has factual errors and does not describe the setup in an orderly way. Worst of all is the use of slangy, flippant style in a very inappropriate situation. There are places, even in formal reports, in which colloquialisms may be employed if care and taste are also used. In this example, the slang is juvenile and offensive to most likely readers. A supervisor reading such a report may feel that he is being made fun of. He would most likely prefer a sense of formality in written reports. Engineers and company executives will regard such a report as the amateurish work of a fool with the sense of propriety of a high school sophomore. It is certainly not a report to be shown to anyone outside of the company as an example of in-house technical writing.

Besides the obvious misuse of English, here are numerous other faults in this bad example. For example, there are 86,400 seconds in a day, so that there will be 172,800 cycles of operation per day. Not a fatal error, perhaps, but one which bespeaks carelessness. In the first paragraph we are left wondering how the number of cycles to failure is recorded, since it is implied that both the relay and the counter are actuated by the cam operated switch. This point is clarified after a fashion in the third paragraph which notes that the counter is actuated by one of the relay contacts, which of course ceases to open and close if the relay itself fails, leaving the counter registering the number of cycles to failure.

Technical writing must find a balance between strict formality, and a pleasant but authoritative style that conveys information smoothly. Our example represents an extreme of informality. Let's try a version that goes well in the other direction, perhaps the kind of report that would be prepared by a testing laboratory for a customer. See the following example.

___

In this test, relay performance was determined by means of life tests made under the following four conditions:

1. Relay at room temperature and humidity (72°F, 60% relative humidity).

2. Relay in a high humidity chamber (98% relative humidity) during test.

3. Relay subjected to saltwater spray for one hour, then tested.

4. Relay tested while exposed to a continuous saltwater spray.

The relay coil was picked by voltage pulses obtained from a cam actuated switch, at a frequency of two cycles per second. One of the relay contacts was used to actuate a counter which recorded the total number of operation cycles until failure. The general arrangements for the test are shown in Fig. 1.

In observing the above forms of this report, note that the number of cycles per day or in ten days is not given, since this information is not properly a part of the test set-up description. The report should not include superfluous material.

Summary

Reports are among the commonest types of technical writing. They range from brief informal memos which supply bits of information for the record, to book-length descriptions of complete projects. Subject matter may be highly technical, or relatively non-technical as in the case of the personal-evaluation example given in this lesson. The intended reader of a report may be one person, with whom the writer has a close and informal relationship, or it may be a person unknown to the writer. Some reports are read by one person only, while others may be printed in four colors and sent to thousands of stockholders of a company.

All of the foregoing factors must be considered in preparing reports, and in deciding on the style of writing and the technical depth that is appropriate. In general, in a decision between informal and formal style, it is best to lean toward the formal when in doubt. When in doubt as to the level of vocabulary to use, it is best to write *slightly* above the assumed level of the intended reader, as this will probably give him a feeling of confidence in the authority of the writer.

It is difficult to provide extensive practice exercises in one lesson of a subject of this nature, since reports generally are based on a much larger volume of data than it is practical to provide here. The best practice will be found in applying the principles here developed in the writing assignments which come as a part of your job.

TEST TC-4

1. Technical reports are usually written in first person...................... _____

2. Fear of writing among engineers probably starts in school.................. _____

3. The engineer who does the work is best qualified to write progress reports.... _____

4. Technical manuals are final reports addressed to the president of a company... _____

5. Most companies do not like to have their engineers write articles for publication... _____

6. Technical proposals are often used by companies to get new business.......... _____

7. Instruction manuals are usually used within the originating company........... _____

8. Technical reports are best written by a professional writer who interviews the engineers who generate the information................................. _____

9. Design notebooks are rarely used in writing progress reports................ _____

10. The Summary of a technical report is placed ahead of the main body of the report _____

MULTIPLE-CHOICE QUESTIONS

11. The first item in a formal technical report is:
    1. Table of contents   3. Title page   5. Index
    2. Summary   4. Letter of transmittal   _____

12. Reference material from which a report is compiled is listed in
    1. The summary   3. The letter of transmittal   5. The contents
    2. The index   4. The bibliography   _____

13. Credits and acknowledgements to persons assisting in the preparation of a report should be given
    1. In the bibliography   4. On a special introductory page
    2. In the summary   5. In the letter of transmittal
    3. In the Exceptions   _____

14. In the sentence, "The electrons had a _____ of 3.4×10^8 cm/sec", the correct word to use in the blank is
    1. Velocity   3. Acceleration   5. Transmission rate
    2. Mass   4. Rate of displacement   _____

15. Which expression used in a technical report would be least likely to sound "out of place"?
   1. Quick as a wink
   2. Sound as a dollar
   3. Red as iron rust
   4. Kneehigh to a grasshopper
   5. Cute as a bugs ear

   _____

16. Which is the best synonym for "surrounding"?
   1. Ambient
   2. Converging
   3. Immersing
   4. Impelling
   5. Sacrocent

   _____

17. Which of the following is the best wording?
   1. On account of parallax, we couldn't get an accurate reading
   2. Because of the angular relationship between pointer and scale an accurate observation was impossible
   3. The pointer and scale didn't line up so the accuracy was all shot
   4. Parallax prevented accurate readings
   5. Due to parallax, the accuracy was no good

   _____

18. Which of the following expressions, used in a technical report, is best?
   1. The frequency ran from 2 cps to 500 megahertz
   2. The frequency range was 2 Hz to 500 MHz
   3. Frequency went from 2 Hz all the way up to 500 MHz
   4. The frequency ran between 2 hertz and 500 megacycles
   5. Frequencies ran from a low of only 2 Hz to a high of 500 MHz

   _____

19. Which of the following best expresses, in a technical report, that a meter "stopped working"?
   1. Conked out
   2. Assumed an inoperative condition
   3. Became unregistrative
   4. Burned out
   5. Failed

   _____

20. The basic conclusion of an investigation report should always be put in the
   1. Summary
   2. Letter of transmittal
   3. Bibliography
   4. Table of contents
   5. Description of test procedure

   _____

LESSON TC-5
Writing Technical Manuals

Introduction

The technical manual differs from the technical report in its purpose and in the reader for whom it is intended. The purpose is to give information and instructions about a piece of equipment, concerning its operation, maintenance, or repair. The intended readers are the users of the equipment, or those who will maintain and repair it. Because of these differences, technical-manual organization is quite different from the organization of a report, and the writing style in a technical manual is directed toward a reader of different vocabulary, interests, and often intellectual level, than those of the executive who typically reads a report.

In this lesson we shall study in detail how these general observations are applied in the preparation of technical manuals.

Technical manuals vary greatly in detail, technical level, and length. At one end of the spectrum, the instructions for appliances and similar consumer items may be given in one page. The typical owners manual for an automobile is only a few pages long and does little more than identify the controls. At the other extreme, the operation and maintenance manual for a ground approach radar system can occupy many volumes. The preparation of major technical manuals can take as long as the design of the equipment they describe; indeed, a frequent problem is found in getting the manual printed before the equipment is obsolete!

Unlike technical reports, manuals are seldom written by the engineers who developed the item. Technical manuals are nearly the exclusive province of professional technical writers. The writers know style and format, and act as reporters in gathering the information that must be included.

The word, manual, comes from the Latin word for hand, manus. The literal meaning of manual is therefore a book that is kept in one hand while a person is learning how to operate a machine -- or at least is kept on hand during the learning process.

Different Kinds of Technical Manuals

Technical manuals provide information to facilitate the use of equipment. In general, they tell:

1. How to operate the equipment.

2. How to adjust, calibrate or otherwise prepare the equipment for use.

3. How to lubricate, clean or otherwise perform maintenance on the equipment.

4. How to test for malfunction, including "preventive maintenance" that seeks to discover potential breakdowns before they occur.

5. How to take apart and repair the equipment, or to replace worn or defective parts.

It is possible to include all of the foregoing functions in one manual, but usually this is not done. A manual given to the user would certainly include information listed under (1), and possibly (2) and (3), but probably would include neither of the discussions dealing with heavy maintenance and repair.

Several manuals are required for a major piece of equipment, such as an automobile. They would include an "Owner's Manual" which tells in non-technical language what the controls are, and how they are operated. A much more detailed service manual tells where to lubricate, how to adjust brakes, timing, breaker point gaps and other parts. A complete manual would explain how to make major repairs, including sequences of operations for taking the car or engine apart.

For relatively simple equipment, such as a military rangefinder or radio, a single manual might be preferred since equipment used far from formal service facilities has to be maintained by the operator. Such a manual might be composed of the following divisions or chapters:

1. General description, including all names and identification numbers.

2. Unpacking instructions, and preparation for use.

3. Instructions for operation.

4. Discussion of the principles of operation.

5. Routine maintenance (by the user, in the field).

6. Overhaul (by service personnel with shop facilities).

7. Parts list.

The order of sections in a manual such as the foregoing is the same as the steps followed by the user on receipt of the equipment. One is tempted to visualize the user with the book in one hand while he unpacks, adjusts and tries out the device with the other hand. The unpacking instructions can be very important, if moving parts are tied down by special bolts; in some cases an attempt by the user to operate without taking all preliminary steps can ruin the equipment.

The discussion of operational principles is of value because such background knowledge elevates the user's understanding above the mere blind following of rules. The operator will always do a better job if he knows "why" as well as "how". Such basic knowledge also helps him to recognize when things are not working properly, so that he can shut down for adjustments or repairs before damage is done.

Some of the most complex manuals are those required for major military systems. The manual for a missile system might consist of <u>ten</u> <u>volumes</u>, for example, covering the following subjects:

1. <u>Basic instructions for operation</u>. This manual takes the user step-by-step through the operation of the system, telling which button to press and which meter to read. It uses non-technical language as much as possible and is addressed to the operating crew.

2. <u>Routine maintenance instructions</u>. This manual is addressed primarily to the operating crew, and describes the checks, test, and adjustments which must be made on a daily basis, as well as lubrication and routine replacement of parts, such as indicator lamps which may burn out, etc.

3. <u>Description of the system</u>. This manual will serve as a general introduction to the system, for both operating and maintenance personnel. In style it may lean somewhat toward the academic textbook. It views and describes the system as a whole rather than from the viewpoint of the dials and buttons, and may include some theory, if this is not too complex. Reading of this book before the first manual mentioned above, will make the latter much easier to follow. Reading of it after the first manual will serve to clarify many instructions the purpose of which may have been obscure on earlier reading.

4. <u>Unpacking and preparation for use</u>. This volume is addressed to either operating or service personnel. It should be complete enough to instruct men unfamiliar with operation in the initial set-up and adjustments needed.

5. <u>Special manual devoted to subsystem</u>. When personnel are assigned to the specialized operation of one part of a large system, they may not need to know in detail how other parts work. Time is saved and confusion avoided by fragmenting the general operating manual. Such subsystem manuals cannot be merely sections extracted from larger books, but generally must be written especially for their specialized readers.

6. <u>Site preparation and repair</u>. A missile system may require the construction of launching pads, silos, and other facilities before the equipment can be installed. Instructions covering this preliminary work are vital if the equipment is to be placed properly. Maintenance and repair of subsidiary structures is also important and should be covered in this manual.

7. <u>Inspection and maintenance</u>. All equipment must have regular inspection and testing, followed by maintenance. Such service is just as necessary for a missile system which is on standby as for equipment in daily use. Inspection, indeed, may have to be much more careful, to detect small deterioration such as corrosion, which could prove disasterous on the day that the system is called

upon to operate.

8. Manuals are often required for subsystems which detail repairs and re-placements most conveniently made at the installation site.

9. Similarly, subsystem manuals are needed covering repairs of a more se-rious nature which require equipment not available at the site so that the equip-ment must be moved to a central shop or depot.

10. Manuals relating to the stocking of spare parts, and matters relating to shipping, storage conditions, and the general maintenance of a backup organiza-tion to support the system.

In cases where parts of systems or subsystems are classified as secret, top secret, etc, the classified details are not given in manuals intended for general distribution, but are contained in special manuals having controlled distribution restricted to personnel with appropriate clearances.

A multi-volume manual of the kind indicated in the foregoing outline would not be prepared by one writer. As many as forty or fifty might be needed to cov-er the range of technical subjects, and to get the books out before the equipment is obsolete! A technical writer familiar with optics, for example, might be as-signed to write on rangefinders or other optical devices. A writer without spe-cialized technical background might be assigned to the preparation of part 10.

Compiling Material

It is a rare technical writer who starts out on a writing job with sufficient knowledge about his subject to just sit down and write. Frequently he may take on an assignment to prepare an authoritative discussion of a piece of equipment he has never heard of. Initial ignorance of the subject is not a matter for shame. On the contrary, it may really constitute an advantage for the writer. If he writes a description in the way that he himself learned, he may have a document that is easier to read than one prepared by an expert so full of details that he doesn't know where to start.

The primary source of information for the manual writer is the designer of the equipment to be described. This information may be obtained by reading technical reports prepared during the progress of the project, or by personal interview with the designer and others concerned with developing and building the equipment. The writer must exercise a degree of diplomacy in conducting interviews, since the latter periods in a development program may be times of stress for the designer -- times at which he is working out bugs in the equip-ment. If the engineer is desperately seeking to make things work right, he will not be in a mood to explain to a writer how often the equipment needs lubrica-tion. Before going to the responsible engineer, the writer should get as much of his information as possible by reading reports and talking to technicians, drafts-man, and others concerned with the project. Often it is possible to write up a preliminary draft of the manual from these other sources of information, and then to submit this draft to the designer who makes his contribution by comments and corrections.

Much material in a manual is most conveniently expressed in the form of tables and charts. Getting this information together is pure compiling and not "writing" at all. The function of the technical writer here is strictly one of getting the information together and putting it in the right place. The detailed steps in making a routine check of a piece of equipment may be much clearer when given in tabular form, rather than in a paragraph of narrative writing. For example, consider the following paragraphs of test instructions:

With the signal generator set to 6.25 MHz, connect the ground side of the output to the chassis and the high side to test point B. Now measure the voltage at TPA, adjusting R7 and L21 alternately for maximum reading. Next do the same thing with R4 and L13, but this time with the voltmeter at TPD and the signal generator adjusted to 7 MHz and connected to TPA. Finally, put the generator back to 6.25 MHz and adjust variable inductance 17 for maximum voltage at TP14.

In tabular form this becomes:

Signal Generator		Voltmeter	Adjustment
Frequency	Connection		
6.25 MHz	TPB	TPA	Adjust R4 for maximum voltmeter reading.
same	same	same	Adjust L21 for maximum voltmeter reading. (Repeat R4 and L21 alternately until maximum possible reading is obtained.)
7 MHz	TPA	TPD	Adjust R4 for maximum voltage.
same	same	same	Adjust L13 for maximum voltage. (Repeat R4 and L13 alternately until maximum possible reading is obtained.)
6.25 MHz	TPR	TP14	Adjust L17 for maximum voltage.

It is clear that the tabular form is much easier for a technician to follow while making the tests. There is one additional piece of information given in the table which is missing in the paragraph -- namely, the fact that the signal generator is connected to Test Point R for the third check. This fact illustrates a fringe benefit is using tabulated instructions. The connection to TPR was accidentally omitted from the paragraph, and this might not have been missed until the manual was published in a way that frustrated technicians. When the instructions were put in tabular form, the blank space in the table was immediately apparent and the writer hurriedly got the necessary information and thankfully put it in.

While the writer is gathering information for the written part of the manual, he is also usually expected to decide what photographs, drawings and diagrams are needed to supplement his text. In making such decisions he should consult with the project engineer and other company officials concerned with manuals, but usually the basic responsibility for selecting illustrations lies with the writer. Photographs are best when made of production items, or at least of prototypes. If these are not available at the time the manual is being prepared, a mockup may be available for photographs. If there is not even a mockup, then it will be necessary to prepare a rendering by the art department. The writer must give the artist the necessary information in the form of blueprints or rough sketches, and see to it that the designer approves of the rendering. Line drawings are usually prepared in the art department from blueprints, and occasionally a blueprint itself may be suitable for inclusion in a manual, although often the blueprint is too detailed for use.

Technical Manual Style

In any discussion of writing we constantly hear the word "style". Style is the selection and use of words and phrases. The writing must be grammatical, but beyond mere correctness there lies the selection of tense and person and the general organization of words and sentences, which is called style. Style may be classified in many different ways. In the previous lesson we compared formal and informal style, the latter being characterized by words such as "we", and the most informal by the direct address of the reader as "you". In general, manuals are more formal in style than reports are.

Many manuals are written for equipment manufactured for the U. S. Government. The government publishes very complete directives or specifications for the manuals which accompany this equipment. Such specifications are usually referred to as "specs". They describe in great detail how manuals should be arranged, what material must be included, how certain words may be abbreviated, how to number paragraphs, how illustrations are numbered and captioned, and finally, what style of writing should be used.

For example, the Air Force states the following in its specs:

"3.7.5.1.1 WORDING OF TEXT. The text shall be factual, specific, concise and clearly worded so as to be readily understandable to relatively inexperienced personnel performing the work on the equipment, yet provide technicians with sufficient information to insure peak performance of the equipment. The sentence form shall be simple and direct, avoiding the elementary and the obvious, and omitting discussions of theory except where necessary for practical understanding and application, or as required by the applicable detailed specification. Engineering knowledge reflected in the manual shall first be converted into the most easily understood wording possible. Technical phraseology requiring a specialized knowledge shall be avoided, except where no other wording will convey the intended meaning. The prime emphasis shall be placed upon the spefic steps to be followed, the results which may be expected or desired, and upon the corrective measures required when such results a re n o t obtained rather than on the theoretical aspects of the work."

The foregoing paragraph is an excellent guide for any technical manual writer. In the case of a manual which forms part of a contract with the government, it becomes a firm directive for the writer. If he ignores such a spec, the government has the legal right to reject the manual, and delay acceptance of the equipment itself.

Here is another example from the same Air Force spec.

3.7.5.1.2 GRAMMATICAL PERSON AND MODE. The second person imperative shall be used for operational procedures; for example, "Break casing head loose from wheel flange". The third person indicative shall be used for description and discussion, for example: "The torsion link assembly transmits torsional loads from the axle to the shock strut."

Analysis of the foregoing paragraphs will show that the government specs are not really the straightjackets that some writers call them. They do impose some restrictions on style, but these restrictions reflect a great deal of experience in writing for the typical technician who must use the manual. Experienced writers do not resent government-writing specs, and they often turn to them f o r guidance in writing manuals for non-government equipment.

In technical writing, clarity of expression takes precedence over variety o r elegance of style. In ordinary descriptive prose, a writer may draw upon the richness of the English language to use synonyms for words in order to avoid repetitions. In a technical manual, however, such practice can cause confusion. If a certain part is referred to as a "top plate", for example, it must always be so designated, and not as a "dust cover" or "circular plate", even though these terms may accurately describe its function and appearance.

Technical writing must usually cram a great deal of exact meaning into a few words. To accomplish this, sentences contain many nouns a n d f e w adjectives and adverbs. For example, the following description of the adjustment procedure for an oscillator is typical of the high information density found in many manuals:

"6.2.7 OSCILLATOR. The oscillator is a variable-frequency electron coupled type, using tube JAN-837 and variable capacitor C1 to obtain the frequency range 300 to 600 kHz. Variable inductor L1 and trimmer capacitor C2 are used to provide adjustment of the resonant frequency when setting the frequency scale attached to C1. When the capacitance of a circuit is small, change in frequency for a given change in C1 is greater than when the capacitance in the circuit is large. Therefore, inductor L1 is adjusted when C1 is maximum (minimum frequency) and trimmer capacitor C2 is adjusted w h e n C1 is minimum (maximum frequency)."

This example states what the technician needs to know. The first sentence states the use of the tube and capacitor C1 in the oscillator. The second sentence tells the function of L1 and C2. The third explains the theoretical basis for the adjustments which are described briefly in the final sentence.

Writing style as compact as this is not for casual reading. It is too condensed even for serious study, as in a textbook or correspondence lesson. It is typical of the high efficiency in style demanded of technical manuals written to give information to technicians and others skilled in the art.

Practice Problems

The best practice in technical manual writing is obtained by writing a technical manual. Many writers learn by doing, getting paid for it in the process, and the results are often very bad. In the following practice problems, the intent is to give you practice in some of the kinds of writing demanded in technical manuals. In order to keep within the space limitations of this lesson, the examples are simple, but they illustrate the principles we have discussed. The "solution" given at the end of this lesson obviously cannot be matched exactly in your writing, but they can serve as examples of adequate answers to the problems presented.

1. Describe in formal terms the small, hand pencil-sharpener of the type usually sold for 15¢ to 25¢. Mention the three parts -- body, blade, and screw -- and indicate how they go together.

2. Most switches for lights in a room are simple off-on switches -- that is, single pole - single throw. It is sometimes desirable to control lights from two places, so that they can be turned on when entering the room on one side and turned off when leaving the room on the other side. Thus if the lights are off, they can be turned on by either switch, and when on they can be turned off by either switch. (If both switches are snapped simultaneously, nothing will happen; the lights will stay as they are, either off or on). Single pole - double throw switches are used for this purpose. Work out how the switches and lamp must be connected to get the effect described, and then describe the circuit in words, without using a circuit diagram, in such a way that an electrician can wire the circuit together by reading your instructions.

3. Describe the basic action of a four-cycle gasoline engine, as one might write it for the introductory section of a service manual on the engine.

4. Describe the action of a photomultiplier vacuum phototube, including a discussion of the circuit consisting of resistors in series connected between the dynodes. If you do not know how this tube works, look it up in a textbook or reference book.

5. Make a freehand sketch of the object described as follows.

The part is composed of three pieces: (1) a plate measuring 12x24 inches, one-half inch thick, (2) a plate measuring 6x24 inches, one-half inch thick, and (3) a 90° angle with equal legs 2 inch wide, 24 inches long, and 3/8 inch thick. The angle is used to attach the two plates along their long sides, using seven equally-spaced rivets through each flange of the angle. The 12 inch plate is butted against the six inch plate. The 12 inch plate has three holes 3 inches in diameter, one centered along the 24 inch length, and the others six inches center-to-center on either side. The centerline of the three holes is midway on the 12 inch dimension. The six inch plate has three rectangular notches measuring 3 inches long by 2 inches deep, cut in the edge opposite the angle, and matching the holes in the 12 inch plate. Both plates have seven $\frac{1}{2}$ inch holes equally spaced, their centers one inch from the edges opposite to the angle.

Make the sketch in perspective, with approximate scale, using a viewpoint which shows all of the features described to best advantage.

6. Give step-by-step instructions for gear-shifting in a standard three-speed stick-shift car, indicating how the shift lever, the clutch, and the throttle are employed in sequence. First describe the procedure for bringing the car from a stop to cruising speed, using the gears. Then explain how to "double clutch" to bring the car from third speed down to second.

7. Rewrite the following passage in a technical manual, correcting spelling, grammar, and any examples of style which you feel are poor.

1-27 ADJUSTMUNTS

a. First you should adjost all the valve clerances in the same order as the firing order of the cylinders.

b. To insure the right clearance, the next thing to do it to back off the clearance adjusting screws several turns or more.

c. Then the cylinder number 11 piston is put exactly on the top dead center for its exhaust stroke.

d. Now we want the cam to be in contact with the cam bearing that is adjacunt with cylinder 11, which is most easily done by the simple process of using the two rocker arm depressers to relieve the value spring load on tappets 7 and 15.

(If you can make clear sense out of d on the first reading you are to be congratulated. However, the information is all there; it just needs to be reorganized.)

8. Rewrite the following technical paragraph to a simpler form, suitable for reading by a technician having little or no knowledge of the mathematics of al-

ternating-current circuits.

The characteristic impedance of a transmission line, defined as $Z_0 = \sqrt{Z_{oc}Z_{sc}}$, is a function of the series inductance per unit length and the shunt capacitance per unit length. When a line is terminated by Z_0, the coefficient of reflection is zero and the transmission line absorbs all power applied to the sending end, since the input impedance is also Z_0. When the terminating impedance has values other than Z_0 there will be reflection so that the line is not capable of absorbing all incident power. On a physical basis, we may say that a short-circuited line is characterized by reflection 180° out of phase with the reflections in an open-circuited line. As the terminating impedance is increased from 0 toward Z_0, the amount of reflection decreases, becoming zero when the termination equals the characteristic impedance. When the terminating impedance decreases from infinity, reflection decreases in a similar manner. Z_0 is therefore seen as an intermediate between Z_{sc} and Z_{oc}, and it is in fact the geometric mean of these as shown by the defining equation.

Outlining and Classifying

We have said very little about the division of a manual into sections and subsections. Textbooks are usually divided into chapters, and perhaps into numbered major divisions within each chapter. Manuals are usually sub-divided in much more detail, since they must be referred to frequently in such a way that a reader can quickly find some specialized description or specification. Standards for subdivision are given in government publications. These are required for manuals written for equipment manufactured for government use, and are usually also followed for commercial manuals since the outlines are logical and are generally well understood by all technical men.

The basic philosophy underlying outlining is simple. The major subdivisions of the work are indicated by distinctive letters or numbers, for example, Roman numerals I, II, III, IV, etc. Within each of these divisions, a number of smaller divisions are indicated by the letters A, B, C, D, etc.

Within each letter division, further division is indicated by Arabic numerals, 1, 2, 3, 4, etc., perhaps printed in boldface type. Further division is shown by lower case letters, a, b, c, d, etc., and by lightly-printed numbers.

A typical outline following the foregoing system might be the following:

Assembly, Aircraft

I. Fuselage

   I-A. Forward section

    I-A-1. Pilots compartment

     I-A-1-a. Instrument panel

and so on

II. Wing

  II-A. Inner wing

   II-A-1. Main shear beam

    II-A-1-a. Upper beam cap

and so on

Each major subdivision contains in <u>its</u> subdivisions all of the associated component parts.

Instead of alternating numerals and letters, outlining can be done with numbers only, using periods between. Thus, a classification VII-D-21-m would become 7.4.21.13 . The all-number scheme is required in some military manuals.

Numerous other systems of classification have been proposed and are used.
One of the most complex is the Dewey Decimal System for classifying books,
used in most libraries. This system undertakes to classify all human knowledge,
and has the built-in capability to extend itself indefinitely into the details of developing subjects. Although the system was designed many years ago, it can be
extended to exactly classify such subjects as field-effect transistors or pulsars.

The purpose of classification is quick reference. If we wish to find a description of a detail or part the location of which we know, we find first the major
section of the equipment in which it is located, then the sub-section, and the sub-
sub-section if necessary until the part desired is located. Detailed classification
also makes it easy to refer exactly to a part of a manual, either in another publication or in a different section of the same manual.

More Practice Problems

9. The following is a list of topics in a manual. Organize these using the
headings I, A, 1, a, etc.

Flying Spot Scanner

1. General Description	6. Connecting cables
2. Parts	7. Unpacking instructions
3. Main cabinet	8. Cathode ray tube
4. Card and slide holder	9. Motion picture attachment
5. Motion picture attachment	10. Power requirements

11. Connection to monitor

12. Adjustments before operation

13. Focus

14. Optical focus

15. Electrical focus

16. Motion picture bypass mirror

17. Operation

18. Loading

19. Loading cards in stack

20. Loading slides

21. Changeover to motion picture oper-
    ation

22. Color balance adjustments

23. Maintenance

24. Cleaning of optical parts

25. Lenses

26. Mirrors

27. Electrical checking of components

28. Photocells

29. Circuitboards

30. Replacement of parts

31. CRT assembly

32. Photocell assembly

33. Lens

34. Stack loading assembly

35. Parts list

The outline for a technical manual should be prepared before any of the manual is written. It then serves as a guide to the writer, as well as a useful aid to the reader.

Each section of a technical manual must be independent, and able to "stand on its own feet". By this is meant that a reader seeking information about a given part or subject must find everything he needs to know in one referenced section, without the need to read everything preceding it. Extensive cross referencing can waste the time of a busy technician (who also may be working under pressure) and may be a source of confusion and error.

10. The following five accounts constitute an exercise in deductive logic and semantics. They are adapted from an exasperating puzzler, told to this writer years ago. Four of the little stories which follow are false, and one is true. The true story can be identified by the way the story is told, from purely logical considerations. Only one story is true, the others are all absolutely false. Which story is true?

a. In a small town in southwestern Missouri there is a drugstore at the corner of Main and Front streets. In front of this store there is a wooden indian, with a bunch of cigars in one hand. Across the street is a fire station, manned by members of the volunteer fire department. Everytime the firebell rings the firemen rush from their homes and places of business, and the wooden indian also jumps off his pedestal and runs across the street and rides out to the fire where he watches it being put out. When this is done, he rides back to the station, runs across the street, and hops back onto his pedestal.

b. There is a small town in southeastern Missouri where there is a drugstore diagonally across the street from the volunteer fire station. In front of the drugstore on a pedestal stands a wooden indian, with a bunch of cigars in one hand. Everytime the indian hears the firebell ring, he jumps off his pedestal and joins the firemen who have rushed from their homes, and runs across the street to ride out to the fire and watch it being put out, after which he rides back, runs across the street and hops back onto his pedestal.

c. In a small town in southeastern Missouri, there is a wooden indian out in front of a drugstore that faces the volunteer fire station. Everytime the firebell rings and the firemen rush from their homes, the wooden indian runs across the street to ride out and watch the fire being put out. He keeps his distance, however, since being made of wood fires worry him. When the fire is out, the indian rides back to the station, jumps off the truck and goes back to his pedestal in front of the drugstore.

d. In a small town in southwestern Missouri there is a drugstore which has a wooden indian in front, a leftover from the turn of the century, which grasps a bunch of cigars in one hand. Everytime the volunteer firemen hear the firebell ring, they rush from their homes, and are joined by the wooden indian who runs across the street to ride out and watch them put out the fire. When this is done, they all ride back together to the station, where the indian gravely stalks back to his position of duty in front of the drugstore.

e. There was once a small town in southeastern Missouri, around the turn of the century, which was noted for a remarkable wooden indian. The peculiarity of this wooden indian, other than his handfull of cigars, was the fact that everytime the bell in the volunteer fire station across the street rang, he jumped off his pedestal and joined the volunteer firemen who rode out to put out the blaze. When the fire was out, this astonishing wooden statue rode back on the truck to the station, where he sedately assumed his position of vantage, without losing one cigar.

General Summary

The writer of a technical manual has a number of functions. He is a complier of information, an organizer of that information into a form permitting easy reference, and a textbook writer when necessary to clarify a matter difficult to understand. In writing sections dealing with maintenance and service adjustments, he must understand clearly what he is describing. A good manual writer must be able and willing to put on coveralls, take tools in hand, and actually go through the adjustment and assembly procedures he is describing. The engineers will of course guide him, but he should be satisfied in his own mind that the procedures they specify are valid, and that he really understands them.

If it is feasable, the writer should have an average technician (as a guinea pig) follow the instructions in his manuscript. Such a trial will often disclose unclear places in the writing, and even errors in the procedures called for by the engineers. The writer recalls a case in which instructors in instrument calibration, in learning the supposedly standardized procedures they were to teach, found that the procedures wouldn't work, and had to help the engineers make corrections!

Manual writing is an art, even though much of the product may appear very uncreative. Technical manuals have little literary merit, but they may contain a greater concentration of exact information than any sonnet does. They do not require nobility of concept and expression, but instead, need careful disciplined thought in planning and execution. In our complex and difficult world there is need for both kinds of writing. These writing styles, incidentally, need not be incompatible. The discipline of accurate technical writing can be helpful to the literary writer, in distilling elegant expressions from the unruly outpouring of pure inspiration. Before you can be a good technical writer, you must be a good writer. When you are the latter, it is easy to take on the specialized skills of manual writing. Literary skill in novels, short stories, or even poems, is really technical in nature. Whether you are describing a sunset or a circuit board, your job is the same -- to create in the mind of the reader a desired thought or impression. If the reader thrills to your sunset, or understands clearly how the circuit board works, you have accomplished your goal; you are a good and effective writer.

Solutions to Practice Problems

The practice problems given in this lesson do not have unique solutions like those in most other portions of the Grantham Engineering Series. The "solutions" which follow are not the best possible solutions. In fact, your solutions to the writing problems proposed here may be superior to our offerings; if they are, we congratulate you and predict success for you in the field of technical writing. The most important element in these "practice problems" is the practice you get in doing them. Here are our "answers".

1. Assembly, pencil sharpener.

The assembly consists of three parts: the body, the cutter, and one machine screw holding these together.

The body is a block of molded urithene, having a conical cavity which guides the pencil, as illustrated in Fig. 1.

The cutter is made of high-carbon steel, of the form shown in Fig. 2. A cutting edge is ground along one side, and a clearance hole is provided for the mounting screw.

The mounting screw is a 4-40 machine screw, $\frac{1}{4}$ inch long, which holds the body and cutter together.

In the foregoing formal description, little information is presented which is useful in manufacturing the pencil sharpener. Reliance must be had in detailed dimensioned blueprints of the body and cutter. Shop information on an item of this kind is usually transmitted by drawings alone. Formal descriptions are not usually given, except for items of greater complexity than our example. In some cases description is needed to provide information not conveniently given in blueprints alone.

2. The circuit requires two single-pole-double-throw switches. Each switch consists of a center contact C which may be connected either to a side contact A or a side contact B. In the complete circuit, one side of the power line goes to contact C_1 of one switch. The other side of the power line goes to the lamp or appliance to be controlled, and then to contact C_2 of the other switch. Contact A_1 connects directly to contact A_2, and contact B_1 connects directly to contact B_2.

Operation of the circuit is as follows: If C_1 connects to A_1, and C_2 connects to A_2, a conducting path is completed from A_1 to A_2 and the lamp is ON. Similarly, if the center contacts both go to their respective B contacts, a path is also established. If one C goes to its A contact, and the other C goes to its B contact, then the circuit is open. It follows that an open circuit can be made closed by changing either switch, and a closed circuit likewise can be made open by changing either switch, thus fulfilling the requirements for the system.

3. A four cycle gasoline engine operates in a cycle consisting of two complete revolutions of the crankshaft, or strokes of the piston. Each stroke consists of an upward and a downward motion of the piston, and these are designated as follows: intake, compression, power, and exhaust strokes.

During the intake stroke, the piston descends. The exhaust valve is closed, and the intake valve is open, allowing a charge of air and gasoline vapor to be drawn into the cylinder. At the end of the intake stroke the intake valve closes.

During the compression stroke, the piston ascends. Both valves are closed, so that the air-gasoline mixture is compressed, increasing pressure by the ratio volumes inside the cylinder at the lower and upper positions of the piston. Shortly before the end of the upward motion the sparkplug fires, igniting the compressed mixture in the cylinder. The piston completes its upward motion to top dead center as the combustion process is initiated.

During the power stroke, the piston descends. Both valves are closed. Combustion of the fuel mixture produces an increased pressure which forces the piston down and supplies to the crankshaft an amount of energy equal to the length of the stroke multiplied by the average pressure within the cylinder.

During the exhaust stroke the piston rises. The intake valve is closed, and the exhaust valve is open, allowing the products of combustion to be driven from the cylinder.

There is one power stroke for each two revolutions of the crankshaft. For automotive use, the minimum number of cylinders is usually four, which gives a power stroke for each half-revolution, providing a reasonably constant flow of power into the clutch and drive system of the car.

4. The photomultiplier cell combines the effects of electron emission by the impact of photons on a photoemissive surface, with the emission of secondary electrons as a result of the impact of high velocity electrons on additional electrodes called dynodes.

The action of the photomultiplier tube commences when an incoming photon strikes the photocathode, made of a metal such as cesium which allows photoelectrons to escape readily.

Assume, for example, that a single photon strikes the cathode, causing a single electron to be emitted. The cathode is maintained at a negative potential with respect to all other electrodes. Typically, the nearest dynode is at a potential of about $+100$ volts with respect to the cathode, and the second dynode is at about $+200$ volts. If ten dynodes are used, the total potential difference is $1,000$ volts.

The electron emitted by the cathode is attracted strongly to the first dynode. Electrostatic shielding prevents it from moving toward any of the other dynodes. Upon collision with the dynode, two or more secondary electrons are emitted. These electrons are attracted to the next dynode, where each one causes two or more electrons to be emitted. For example, if each impinging electron causes two secondary electrons, there will be four electrons emitted by the second dynode. If the ratio of primary to secondary electrons is 3, there will be nine emitted by the second dynode.

The foregoing process of multiplication continues around the dynodes. For ten dynodes, a ratio of 2 will give 2^{10}, or 1024, electrons per photon falling upon the final anode of the tube. A ratio of 3 will give 3^{10}, or $59,049$, electrons per photon. The current multiplication is greatly influenced by the dynode voltage, since this determines the ratio of primary to secondary electrons at each stage.

The foregoing description is typical of what might be found in a manual. The sensitivity of the tube can be dramatically illustrated, however, by a simple quantitative calculation which is easily made, and which will increase the interest and value of any manual describing equipment in which the photomultiplier is an important part. Consider the following addition as something beyond what is normally expected in a manual, but which will be appreciated by every reader.

The sensitivity of the photomultiplier can be appreciated from the following example: The sky on a clear moonless night has a brightness of about 4×10^{-5} candles per square foot. Light reflected from the ground under such a sky might have a brightness of about 10^{-6} candles per square foot. A square foot of ground would therefore be of 10^{-6} candlepower. Since one candlepower is equal to 4π lumens of light flux, the total light emitted over a hemisphere is 4π x 10^{-6} lumens. Now consider a lens of one inch diameter placed 100 inches from this square foot, which collects light and transmits it to the cathode of the multiplier phototube. The area of the lens is 0.785 square inches, which is 1.25 x 10^{-5} of the hemisphere. Energy calculations show that one lumen of light is equivalent to 16,100 ergs per second. In the middle of the visible spectrum, one photon has an energy of 3.52 x 10^{-10} ergs. Dividing, one lumen is found to correspond to 4.58 x 10^{13} photons, or 45,800 billions of photons !

The 4π x 10^{-6} lumens emitted into a hemisphere corresponds to 5.67 x 10^8 photons per second, and the light entering the photocell is
$$(5.76 \text{ x } 10^8)(1.25 \text{ x } 10^{-5}) = 7,200 \text{ photons per second}.$$

If each photon causes one photoelectron to be emitted by the cathode, and the multiplication factor of the tube is 59,000 electrons per photon, the current at the anode of the photomultiplier will be 425 million electrons per second, or 0.000068 microamperes. A current this small can be amplified by special electrometer circuits, so that the application of a photomultiplier in detecting the brightness of dark ground on a moonless night can be regarded as representing ultimate light sensitivity. Such sensitivity corresponds to what a thoroughly dark-adapted eye can detect.

5.

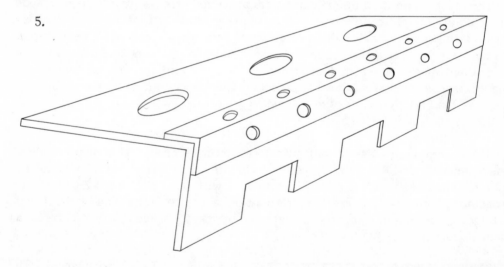

6. Gear-shifting instructions. First, instructions for bringing a car from a stop to cruising speed:

A. Depress the clutch pedal.

B. Move shift lever to LOW GEAR position.

C. Depress foot throttle slightly, increasing engine speed to a point corresponding to a road speed of about 20 miles per hour.

D. Slowly raise clutch pedal.

E. When clutch starts to engage, engine will be slowed down as car starts to move. Depress throttle some more to prevent engine from stalling. Note: The proper coordination between clutch and throttle, when the car starts to move in low gear, is the chief source of difficulty to beginning drivers. If the engine speed is too low, the engine will be killed. If the engine speed is too high, the car will commence moving with a series of jerks. The key to success in this action is to open the throttle just the right amount at the moment that the clutch starts to engage. The exact amount of extra throttle can be determined only by practice, and varies somewhat from one car to another.

F. When clutch is fully engaged, accelerate the car in low gear to a speed of about 10 mph. Release the throttle fully.

G. Depress the clutch and immediately shift the gear to neutral.

H. Pause for about one second and then shift to the second gear position. Note: The pause is to allow the engine speed to decrease from that corresponding to 10 mph in low gear to that corresponding to 10 mph in second gear. In synchromesh transmissions (found in all modern gearshift cars) the synchronizing clutches will be better able to bring the gears to engine speed when the shift to second is made deliberately.

I. Engage the clutch and at the same time open the throttle until the engine pulls the car causing continued acceleration. Accelerate the car until a speed of about 25 mph is reached.

J. Depress the clutch and shift to neutral. Release throttle.

K. Pause for about one second and shift to high gear.

L. Engage the clutch as in Step I, and open the throttle so that acceleration of the car continues.

Now, instructions for down-shifting of gears:

In normal driving the brake is used to reduce speed, or to control speed on a down-grade. When descending very steep or long down-grades, however, use of the brake will cause excessive wear and heating. In extreme cases the heating may be so great as to reduce braking ability. There have been cases of heavy trucks using the brakes to the point that braking action is no longer sufficient to prevent a runaway on a long downgrade.

A necessary part of driving skill is the ability to <u>down-shift</u>. Downshifting is relatively easy with sychromesh transmissions. On some truck transmissions which do not have synchromesh, considerable skill is required to avoid raking the gear teeth together, creating noise and possible damage. Also, if the technique required for non-synchro transmissions is used with the synchro type, the operation will be much smoother and will avoid jolts to the car or truck. The following instructions are written for non-synchro downshifting, but should be followed for both types of transmission. The instructions begin with the assumption that the vehicle is in high gear.

A. Depress clutch and at the same time release throttle.

B. Shift from high to neutral.

C. With gear in neutral, engage clutch and depress throttle slightly to bring engine speed up to a speed corresponding to the car speed, for second gear.

D. Depress clutch and quickly shift to second gear. If the engine speed is correct to correspond to car speed, the gears will engage very quietly. If not, there will be a clash of gear teeth. If this occurs, do not attempt to force the gears together, but depress the clutch (with the gears in neutral) and again bring engine speed up to what you believe to be the correct value. Several tries can be made quite quickly. With experience, it is possible to bring the two pairs of gear teeth lightly together so that the contact between teeth can be felt through the shifting lever. As the engine is accelerated, the difference in speed is sensed as a vibration which gradually slows down. Then the vibration speeds up as the engine speed becomes too high. The skillful driver can feel the instant of synchronism when the vibration ceases, and then quickly depress the clutch and engage the gears. (In some cars, the gears can even be engaged without use of the clutch, although this practice is not recommended.)

E. After the car has run in second gear for a time, and probably slowed down due to engine compression, disengage the clutch again, and speed up the engine with the gears in neutral. Repeat Step D in shifting to low gear.

7. ADJUSTMENTS is misspelled. In (a), "adjust" and "clearance" are misspelled. Also, "you should" is best eliminated. The 'you' should not be used. In (b), the expression "the next thing to do" is unnecessary and may be too informal and the "or more" is superfluous. In (c), the word "then" might be appropriate to a narrative description, but when the outline form is used it is not needed. The trouble in (d) is mainly that the directions are given rather casually at the end of a long involved sentence that starts with several cardinal errors: "Now" is unnecessary here, and the use of "we" is inconsistant with the "person" used in a, b, and c. Spelling of "adjacent" is incorrect, and the homely "simple process" sounds almost like country dialect. As in all instructions, the procedure should be described first, with reasons given later.

The following represents a "cleaned up" version of the poor example, as it might be done by a long suffering technical editor, just before recommending that the writer be fired.

1-27 ADJUSTMENTS

a. Adjust valve clearances in the same order as the firing order of cylinders.

b. Back off valve-clearance adjusting screws several turns to insure proper clearance.

c. Place piston of number 11 cylinder on exact top dead center of exhaust stroke.

d. Relieve valve spring load on tappets 7 and 15, using two rocker arm depressors. This allows the cam to slide over so that it is in contact with cam bearing adjacent to cylinder 11.

8. In the revised version of this technical description, the mathematical definition of Z_O must be omitted, and also such expressions as "coefficient of reflection", "terminating impedance" and "geometric mean". A good way to get the idea over is to use a physical analogue that provides a visual image, and to then describe the situation somewhat informally. We have stressed the use of <u>formal</u> language in manuals, but in a situation in which the reader may be frightened by technicalities, the formality rule may be relaxed a bit in order to instill confidence. There are many ways in which the example given can be simplified. Here is one:

Electric waves on a transmission line act in many ways like water waves in a long trough. If there is a solid wall at the end of the trough, the waves will be reflected. If there is a flexible rubber wall at the end of the trough, with air on the other side, there will also be reflection. The incident waves and the reflected waves act together to form standing waves, which rise and fall in the same positions. The positions at which the peaks and valleys occur do not move. The position of a standing wave caused by reflection at a solid, fixed wall is a half wavelength removed from the standing wave resulting from the flexible wall. In the electrical transmission line, the solid wall in the water trough corresponds to a open circuit at the end of the line and the rubber wall corresponds to a short circuit, if we let the water wave correspond to electrical voltage. Electrical standing waves are formed in both cases, but in different positions along the line.

When there is reflection, the power put into a transmission line is returned; we say that the line will not accept power. Since the purpose of a transmission line is to accept power and transmit it, it is undesirable to have reflection. Therefore we want to use an impedance at the end of a line which will let the line absorb power without reflection. The impedance which allows maximum power absorption is called the characteristic impedance of the transmission line. The characteristic impedance depends on the wire size and the spacing between the wires. If we measure the input impedance of a given transmission line when the far end is open circuited, and also the impedance when the far end is short circuited, the characterisitic impedance is the <u>square root of the product</u> of these impedances, which is called mathematically the "geometric mean" of the two impedances.

In the water-wave analogy to a transmission line, we can replace the vertical wall or rubber membrane by a sloping surface like a beach. The waves will break on this surface and dissipate their energy in turbulence. When the slope of the beach is just right, there will be no reflection, and so we may say that the water trough is "terminated" in its characteristic impedance. Most ocean beaches adjust their slopes automatically to a condition of no reflection, i. e. to a "characteristic impedance" of the transmission medium (the ocean water surface and floor) through which the wave energy travels.

The foregoing example goes considerably beyond the usual limits of a manual, and becomes a small textbook. Such deviations from manual style are occasionally justified when it is necessary that non-engineering personnel understand some theoretical matter in order to do their job.

9. In converting this list of topics into an outline, we first look for the major sections, which will be listed with Roman numerals. These are clearly

I.   General Description

II.   Unpacking Instructions

III. Adjustments Before Operation

IV. Operation

V.   Maintenance

VI. Parts List

Some writers might separate Replacement of Parts from Maintenance; this depends on whether the same person is likely to make the routine adjustments and clean the optics, as well as the more serious jobs of replacing major assemblies. In the case of a complex electronic system such as the <u>scanner</u>, the general manual makes no effort to discuss electronic troubleshooting or major repairs; these are jobs for highly skilled personnel.

Subheadings would be as follows:

I. General Description

  I A Parts

    I A 1 Main cabinet

      I A 1 a Card and slide holder

    I A 2 Motion picture attachment

    I A 3 Connecting cables

II. Unpacking Instructions

II A Cathode ray tube

II B Motion picture attachment

II C Power Requirements

II D Connection to monitor

III. Adjustments Before Operation

III A Focus

III A 1 Optical focus

III A 2 Electrical Focus

III B Motion picture bypass mirror

IV. Operation

IV A Loading

IV A 1 Cards

IV A 2 Slides

IV B Changeover to motion picture operation

IV C Color balance adjustments

V. Maintenance

V A Cleaning of Optical Parts

V A 1 Lenses

V A 2 Mirrors

V B Electrical checking of components

V B 1 Photocells

V B 2 Cathode ray tube

V B 3 Circuitboards

V C Replacements of Parts

V C 1 CRT Assembly

V C 2 Photocell assembly

V C 2 a Photocells

V C 3 Circuitboards

V C 4 Lens

V C 5 Stack loading assembly

VI.   Parts List

10. Story b is true. There are many small differences between the ways the stories are told. Most of these variations have no bearing at all on the logical truth of the stories. You will note that versions a, c, d, and e all say, in one way or another, that everytime the firebell rings the indian runs across the street to ride to the fire and to do other things which no wooden statue could possibly do. In version b the key words are, "Every time the indian hears the firebell ring . . . " Thus he does all the running around only when he hears the firebell. But a wooden indian can't hear anymore than he can run around. Therefore he never hears the bell and hence never does any of the other things, so that the story, b, is true. All the others are false because they do not apply the condition that the indian acts only when he hears, which is never.

If you didn't get this little puzzle, do not be disheartened. In years of trying it out on scores of people, including some distinguished PhD's on university faculties, the writer told it to only two or three who found the key. When the story is told orally, only two versions need be used. You can make them up extemporaneously, putting in as many "red herrings" as you wish, but being careful to have "everytime the bell rings" in one, and "everytime the indian hears the bell ring" in the other.

TEST TC-5

<u>TRUE-FALSE QUESTIONS</u>

1. Technical manuals are written for operating personnel....................... _____

2. Manuals are usually written by the design engineer.......................... _____

3. Manuals usually do not discuss preventive maintenance...................... _____

4. Unpacking instructions should precede operation instructions.................. _____

5. The outline for a manual is prepared after the writing is complete............ _____

6. Most manuals employ a more formal style of writing than do reports.......... _____

7. Manuals covering technical subjects should employ advanced technical wording. _____

8. Outlining of manuals to military specs is left to the good judgment of the writer. _____

9. Pictures and diagrams are usually employed as required by the writer......... _____

10. Good technical manuals are often characterized by a high density of necessary
    information... _____

<u>MULTIPLE-CHOICE QUESTIONS</u>

11. Which of the following phrases is most suitable for use in a technical manual?
    1. Capacitor C3 regulates the frequency
    2. The purpose of capacitor C3 is to regulate what the frequency is
    3. Condenser C3 controls adjustments of the frequency changes
    4. The condenser at point C3 determines the frequency
    5. The capacitor found at position C3 has a value which determines the fre-
        quency   _____

12. Which of the following is best for use in a technical manual?
    1. To raise the frequency, you turn screw B clockwise
    2. When one wishes to increase frequency, he turns screw B clockwise
    3. The frequency is raised by means of screw B which is turned clockwise
    4. To raise frequency, turn screw B clockwise
    5. In order to raise the frequency, it is necessary to turn screw B clockwise _____

13. Which one of the five sketches below best represents the following description?
    "The channel measures 4 inches by ten inches long, with one-inch flanges. Quarter-
    inch holes are drilled two inches from each end of the flat, and four eighth-inch holes
    are equally spaced, $\frac{1}{2}$ inch from the edge of the flange. Exterior flange angles are 90°
    and 120°, on 0.040 sheet stock."

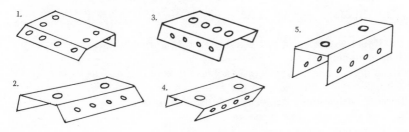

14. Which one of the following descriptions best matches the sketch?

   1. A cube, with four holes drilled thru each corner and one large hole thru the center, and another hole intersecting the large hole

   2. A block, 4 inches square and two inches thick, with a hole of two-inch radius thru the large face, a hole of $\frac{1}{2}$ inch radius intersecting it at 90°, and four $\frac{1}{4}$ inch holes placed 1 inch from each edge at the corners and drilled parallel to the axis of the large hole

   3. A block, four inches square and $\frac{1}{2}$ inch thick, has a 2 inch diameter hole thru the four-inch face, four $\frac{1}{4}$ inch holes 1 inch from the edges at the four corners, and a 1 inch diameter hole passing thru the large hole at right angles

   4. A block, four inches square and 2 inches thick, having a 2 inch diameter hole centered on the large face, with four $\frac{1}{4}$ inch diameter holes 1 inch from the edges at the four corners, drilled parallel to the axis of the large hole, and a hole 1 inch in diameter drilled perpendicular to the 2x4 inch face so that its axis intersects that of the large hole at 90°.

   5. A block $2x2x\frac{1}{2}$ inch, with a 1 inch diameter hole parallel to the large face, four 1/8th inch holes with axes parallel to the large hole, and two $\frac{1}{4}$ inch holes perpendicular to the 2x2 inch face.   _____

15. Which of the following expressions is acceptable in a technical manual?

   1. You should not charge the battery longer than 24 hours

   2. One should not charge the battery longer than 24 hours

   3. Twenty four hours represents the maximum time that the battery should be allowed to remain on charge

   4. The battery should be on charge not longer than 24 hours

   5. The battery should be charged for a period of time not to exceed 24 hours   _____

16. A topic in an all-number outline is given as 3.11.7.9 . This is equivalent to

   1. C.11.-7-j   3. III-K-7-i   5. III-XI-VII-IX

   2. III-11-g-i   4. 3.k.7.i   _____

17. The topics covered in a manual on a TV receiver are as follows: 1. general description 2. electronics 3. RF section 4. first detector 5. second detector 6. video section 7. color phase detector 8. audio section 9. power amplifier 10. tone control 11. crossover network 12. sweep circuits 13. horizontal 14. vertical 15. picture tube 16. power supply 17. cabinet. The number of <u>primary divisions</u> is

   1. 1   2. 2   3. 3   4. 4   5. 5   _____

18. The outline classification of the topic "first detector" in the list in Question 17 would be

   1. 1.2.2   2. 3.1   3. 2.1.1   4. 2.1.2   5. I-A-4   _____

19. Referring to the solution for Practice Problem 9 in Lesson 904, if an all-number system were used for the outline the section describing the optical focus adjustment would be classified as

   1. 3.1.1   2. III.A.1   3. 3.A2   4. 3.2.2   5. 2.3.1   _____

20. A football-shaped object with a steel rod passing thru it so that it goes thru the two pointed ends might be described as

   1. An oblate ellipsoid with a rod passing coinincident to the major axis

   2. A spheroid with a rod perpendicular to the minimum diameter

   3. A prolate ellipsoid with a rod coincident with the major axis

   4. An oval with a rod perpendicular to the short axis

   5. An elongated sphere with a rod along the short axis   _____

LESSON TC-6
Writing Technical Articles

Introduction

The technical article differs from other kinds of technical writing in that it is usually not written as a definite job assignment. Technical articles are written for a variety of publications, ranging from house organs (company magazines or newsletters) to such nationally circulated magazines as <u>Popular Mechanics</u> and <u>Scientific American</u>.

Writers are motivated to write technical articles either by the personal reputation it may bring or by the money to be received for the articles. Reputation may be with the general public, with a specialized professional group, or with the management of the company for whom the writer works. A rather special category of employer consists of colleges and universities, where faculty members are promoted largely on the basis of the number of published articles bearing their names. In such schools, faculty members wryly say that the policy is "publish or perish".

Technical writers branch out into technical <u>article</u> writing in much the same way that newspaper reporters try their hands at fiction writing. If you are a "real pro" in writing, you will find that eight hours a day on the job is not enough, and you will want to express yourself free of the bonds of company standards and customer specs on format and style. If you feel this way, then you should get the old typewriter going. There's nothing quite like seeing your name in print, and when a check comes in for a bit of your moonlighting, the pleasure will be compounded. You will also find, as a fringe benefit, that writing articles on the outside will improve your ability on the job, which will be expressed in pay raises and increased respect from your fellow-writers and boss.

The foregoing rather rosy description of technical writing should be tempered by pointing out that the pay for such articles is not as high as it is for fiction or general-interest articles. Many professional and trade journals do not pay at all, on the theory that writers receive their living from their primary jobs, and write in order to disseminate knowledge they have gained. Even distinguished scientists are invited to contribute articles free, perhaps on the grounds that such exalted people would be insulted to receive a lowly check.

However you choose to go about it, the writing of technical articles for publication will be a fascinating and rewarding experience. If the reward is not great riches, it will nevertheless be substantial in pleasure, recognition, and the general upgrading of your ability as a technical writer.

Where are Technical Articles Published?

The world today is more technically minded than ever before in history. Scientists need to communicate their ideas, engineers to learn about new discov-

eries which they can apply, and technicians to find out what the state of the art
is in their fields. Even the layman, once rather disinterested in esoteric things
like vacuum tubes, is now avidly interested in learning about lasers. All of this
interest has produced a torrent of technical writing. Topics range from the most
intricate scientific subjects to do-it-yourself bookcases, with writers recruited
from university faculties and basement hobbyists.

Throughout the world there are more than 50,000 technical and scientific
journals, in which over a million articles are published each year. About 60,000
new books on science and technology are published each year, and hundreds of
thousands of research reports are printed with distributions running from a few
dozen to hundreds or thousands of readers. This flood of publication is increasing every year and there is no sign that either the amount or the rate of increase
will diminish. The world in in the midst of an "information explosion", and
technical articles are a major part of this.

Technical publications can be classified in several ways. One classification
is according to subject field. There are hundreds of journals covering physics,
chemistry, geology, and the biological sciences, more hundreds in the various
fields of engineering, medicine, and psychology. The merging of scientific dis-
ciplines has produced journals having hyphenated and compounded names to indicate the mixture: Journal of Chemical-Biology, Astrophysics, etc.

Within each subject field we can classify journals according to technical level. At the top are the magazines written by and for the leading research scientists, such as the Physical Review. In mid-range are magazines for the working
practitioner, who cares more for applications than for the rigorous theory.
Slightly below this stands Scientific American, whose goal is to explain to experts in one field what is happening in other fields, and of course to educate the
intelligent layman. Next we have the "popular" magazines: Popular Science,
Popular Electronics, Popular Mechanics (the grand-daddy of them all) and the
semi-technical magazines, such as Popular Photography and Stereo Review.

As the technical level goes down, the circulation of the magazines increases,
and also the payment made for articles becomes better. A check to a contribu-
tor to the Reviews of Modern Physics would be an insult, but the Popular Science
type magazines would have little material if they did not pay for it.

Today, newspapers often accept science pieces. The big metropolitan dailies
employ full-time science editors who do most of their writing, but smaller papers will buy free-lance articles, particularly if they deal with scientific events
of current interest, such as space exploration or landings on the moon.

Many smaller papers publish science columns that are syndicated. The pay
depends on the circulation of the paper, and is in a range of a dollar-per-column
up -- rates of course far too low to attract a writer, were it not for the fact that
a successful syndicated column may sell to hundreds of papers. We shall say
more about syndicating later in this lesson.

Who Writes Technical Articles

Many technical articles are written by full-time technical writers. Even more are written by engineers, technicians, and others who do writing only occasionally. The motivation for such non-professionals is varied. They may write for the pay, as a kind of literary moonlighting, but more often the reason lies in the prestige gained by appearing in print. When one writes on a subject, he may be presumed to be an authority on the matter, to whom his colleagues turn for information or advice. If you are invited to submit an article to a trade or professional journal, you may find that your "status" will begin to improve with your employer. Most engineering firms actively encourage their employees to write technical articles because of the prestige which employers (as well as employees) gain from publication by employees. They will help in such ways as preparing illustrations and editing, and may even assist in the writing itself. Some companies will go as far as letting an employee do some of his writing on company time.

Publication is very valuable to the professional man who is self-employed, since articles serve as advertisements of his competence. Reprints of articles make excellent additions to a resume. Even medical doctors find that there are advantages from writing -- for example, referrals from other doctors, and being called in for consultations.

The power of publication in building up reputation is illustrated in a bizarre way in an area of southern California where a colony of astrologers and "experts" on the occult has become established. The writer of this lesson once had the opportunity of talking very freely with a gentleman who was an acknowledged authority on esoteric mysticism, the author of several books and many articles, and the leader of a devout (and dues paying) gaggle* of followers. He gave this writer the formula for success and profit in the mystic world. It is as follows:

1. Write a book on some appealing mystical idea that you can easily invent. Publish the book at your expense.

2. Advertise the book in one of the many magazines devoted to occult matters.

3. In return for paying for a good display ad, the magazine will write a book review complimenting you highly.

4. People will start sending in orders for your book.

5. People will write "letters to the editor" telling how good your book is, and these letters will be published so long as you keep on buying advertising space in the magazine. (You can write the letters yourself, but you'll be surprised at how many people there are who'll do it for you.)

If you're lucky, you can make it to the next stage, which is to become a full-fledged prophet and religious founder. You then hold meetings, preach sermons, and sell franchises for branch churches like fried-chicken restaurants. More

---

*"Gaggle" is a collective noun meaning group, used exclusively for geese.

than one successful religion has been founded by following these simple rules.

Now, let us return to the more serious subject of writing legitimate technical articles.

The style and content of a technical article depends, as has been noted, on the type of magazine it is written for. Let us consider several kinds of publications and the kinds of articles which they use.

Writing for Technical Journals

The first requirement for the writer of an article for a technical journal is that he be a specialist in the field of the journal. Journals are read by technical people who know the field, and want to learn what is new and advanced. If you have made a discovery or invention (and it is safe to disclose information from the standpoint of patent protection) then you can gain recognition, and by publication perhaps attract users of your invention.

If you are working with someone who had made a discovery, you may be able to get him to agree to your writing an article about his work. You may do this as a pure ghost writer in his name, or you may publish under joint authorship, or you may write it up under your by-line as a report on the work of the inventor. In large companies, the writing of an important research article may be by a technical writer acting as a "ghost", in order to conserve the time of the discoverer, which is better spent on technical work instead of writing. In such a case, if you are chosen to do the writing you will gather information in an interview, and by reading technical reports will write your article for review by the inventor, and after making any tehcnical corrections will prepare the final draft for publication. In this kind of writing, for which you receive your regular salary, you get generally no recognition in a technical sense, since you are a reporter.

As a technical writer for publication, you first place yourself in the position of the reader and then organize and present the material in a form which will be pleasant and easy for him to absorb. You must decide what to include and what details to leave out, either for reasons of industrial security or because they are beyond the technical background of the average reader. Within each of the many specialized fields represented by journals, there are several magazines representing various levels of technical sophistication. An article prepared for one level would be inappropriate for a magazine at another level, either higher or lower.

New discoveries of importance are usually described several times in journals of descending technical sophistication. For example, a basic discovery in physics having potential application to transistors might be first described in Physics Today, in a highly theoretical article prepared for Ph D's who are the peers of the discoverer. Some time later, the same subject might be covered in an article in Electronics or Scientific American. In Electronics, complex theory would be deleted and the discussion directed to the backgrounds of engineers and technicians. In <u>Scientific American</u>, the treatment would be more philosophical, and would seek to describe the fundamental physics of the matter for sophisticated laymen rather than for applications-minded technicians. If the discovery finds practical application, in a new transistor, for example, this news will likely be first reported in <u>Proceedings of the Institute of Electrical and Electronics Engineers</u> at the theoretical level, and in <u>Popular Electronics</u> in a do-it-yourself article for hobbyists and technicians.

All of the foregoing levels of sophistication offer opportunities for the technical-article writer. At the top, he must either be the discoverer or be working closely with him. At the <u>Popular-Electronics</u> level, all you need do to qualify is to buy one of the gadgets and figure out a use for it that will interest other home experimenters.

Some professional magazines publish articles only by invitation, extended to select well-known authorities; others will consider unsolicited offerings. The commerically-oriented magazines like <u>Popular Mechanics</u> will always consider unsolicited manuscripts.

Publication of articles about a new product is an important part of the sales promotion of a company. If you are a technical writer who has prepared reports, proposals, or manuals about a product, you may be asked to write an article about it, to be published under your own byline, and you may even receive a fee from the magazine in addition to your regular salary while writing.

In writing for a technical journal, the writer usually has to start by familiarizing himself with the subject field. No one person can be expected to approach every job fully armed with background information, and lack of knowledge about the subject should not be an embarrassment. The first few days of a new assignment should be devoted to familiarization, by reading basic text material, other articles, and reports dealing with the matter to be described. The writer should also get copies of the magazine for which the article will be prepared, to study its style, format, and level of sophistication.

Once the writer has sufficient general background in the subject, he must obtain from the originator the specific information to be conveyed in the future article. This information is usually found in reports, or in previously written articles at a more sophisticated level, and the mere unwinding of such material may be a real job for any writer. For example, here is an excerpt from a discussion of network design taken from <u>Electro Technology</u>.

> Any three- or four-terminal, active or passive network which is linear (or, if non-linear, which may be treated as linear by consideration of small signals only), may be completely described by a set of four network parameters. These parameters are the coefficients of a pair of equations which relate the input and output voltages of the circuit.

In scaling down this paragraph for comprehension by less technical readers, the writer must of course understand what it means. Then he must consider what technical words can be used without definition, and which must be explained or replaced by more familiar terms. Should the words "active" and "passive" be expanded by saying, for example, "network containing transistors" and "network containing only resistance and capacitive elements"? Will the very useful word "parameter" be understood by the intended readers, or should a less elegant term, like "numbers", be used?

In adjusting the technical level of an article, there may develop a tug-of-war between the writer and the originator, who may instinctively want to describe every feature of his brain child, as a matter of pride. As in all collaborations, patience and diplomacy are necessary to achieve the goals desired by all parties concerned.

The complete theory of a device may be too complex for the writer to undertake or understand. In such a case, the complete treatment must be done by the specialist. A simplified description can usually be written by a writer ignorant of the theoretical foundation, if the expert is available to answer questions and read over the manuscript for errors of fact. Even when the writer does not follow the thread of a technical argument, he can often perform a valuable service in correcting grammatical errors. Sentences are often difficult to understand because of awkward construction. If the writer can rebuild phraseology by seeing to it that every sentence has a subject and predicate, and that pronouns refer to the correct nouns, he can make a complex discussion much clearer to the trained reader, when he (the writer) may not really understand the matter at all!

A frequent source of confusion in writing is the use of the word "this" as a subject, referring to something in an earlier sentence. For example, we might have the following:

The designer should realize that the calculation of centrifugal loads must precede any detailed design of the rotor hub, as catastrophic failure can result from fatigue effects. This is very important.

The word "this" could refer to the importance of failure, the order of calculation and design, the size of centrifugal loads, or the realization of the foregoing on the part of the designer. As written, the second sentence in the example is meaningless. The trouble is easily cleared up by using the appropriate noun, e. g. "This realization is very important", which gives point to the whole thing.

The example given on the discussion of a network is difficult reading, even for the trained specialist. If the writer can make a technical description more palatable, he should do so, no matter how sophisticated the reader may be. Here is a brief excerpt from an article in Electrical Engineering, the journal of the professional electrical engineering society.

Before examining the potential characteristics of the thermionic converter, a quick review of its basic design and a description of 'how it works' are perhaps in order. The thermionic diode basic design is quite simple in concept. It is composed of a cathode (emitter) and an anode (collector). The cathode. . . ."

The reader who understands how a thermionic converter works can skim over this section, since he has been tipped off that a "quick review" is coming up. The transistor terms "emitter" and "collector" are inserted to help older engineers who got their training in vacuum-tube days and may have to pause to relate the older terms like cathode to a modern discussion. Even the most sophisticated reader is also a human being, who enjoys reading more when it does not blast him with a formal description in patent-claim style. (Incidentally, the condensed wording used in claims, so necessary to cover all legal loopholes, provides the world's finest example of how not to write a technical article.)

In his collaboration with the expert, the writer can make a major contribution because of his lack of knowledge of the subject. This apparent contradiction is easily explained. For the writer to understand, he must mentally go through the same stages of learning that must be followed by the reader of the proposed article. If ideas are developed in an order easy for the writer to follow, they will likewise be easy for the reader. In this sense the writer serves as "devil's advocate" in testing out the effectiveness of a presentation, or perhaps the name "guinea pig" is more appropriate.

A skilled writer has developed a sense of order in putting words and sentences together to tell a story. Even though the writer may not grasp every subtle detail, he can still perform a valuable service in smoothing the overall narrative. Narrative continuity is as important in a technical article as it is in a piece of fictional writing. In this function, the writer can often spot an omission that the expert missed because of his superior knowledge.

In the previous lesson we presented an outline for the preparation of a technical proposal. The same basic principles of logical development apply in the writing of a technical article. The first paragraph should contain a brief summary of the fundamental arguments of the article. Like a newspaper story, the article should provide the basic setting for what is to come. Following paragraphs should present theory, experiments based on the theory, and conclusions, in a smooth sequence that builds up ideas in the reader's mind.

The ideal method of ghost-writing an article is for the writer to work up a skeleton outline of the article after preliminary talks with the expert. If the latter has already put many words on paper, the writer's problem becomes more complicated (and delicate, since even the most inept writer loves his brain children). Sometimes, when the original draft is too bad, there is no recourse to a fresh start. More often, however, the writer can make miraculous improvements by just cutting up the original version with scissors and reassembling paragraphs in different order, with perhaps a few bridges of his own composition in between. By this technique, the expert is made to feel that his words are still used. In fact, when he finds how much clearer they are in a different order, he will probably be even more proud of himself. The writer who can foster such beliefs without hurting his own ego, is a pearl beyond price.

A special problem for the writer is created by the new words that are coined to describe new discoveries, and the changes in the meanings of old words made by experts. The problem is not only for the writer to understand the words, but for him to assure himself that the meanings are understood throughout the profession and are not peculiar to one laboratory or even one small group of experts. New technical words are often created jokingly, and when they come into serious use, their inventors are surprised and sometimes dismayed. Eccles and Jordan called their two-state circuit a bistable multivibrator, but users quickly substituted the shorter and more graphic name "flipflop". In nuclear physics, a unit of area used to measure nuclear absorption cross-sections has puzzled the younger generation of physicists. This is the "barn", a very tiny area of 10^{-24} square centimeters. The word came about when Robert Oppenheimer, commenting on the difficulty of hitting such a small target with a nucleon, said it was like hitting the side of a barn.

In writing an article for a general audience, new words should be avoided, unless they are defined in the article. The writer should be a conservative influence, and allow new words to enter only when standard English is inadequate. There are at least three good reasons for such opposition to "progress". (1) The meanings of new words may not be the same in different parts of the country, (2) words not yet found in standard dictionaries cannot be looked up by a reader unfamiliar with them, and (3) the general fact that the use of too many technical terms, new or old, will decrease the overall level of comprehension.

The pruning of fancy language is made more difficult when the expert is the inventor of the word in question. A real problem in tact is presented to the writer in persuading his expert to avoid a ponderous style. He can usually do this by explaining that it is better to be known for clear explanations than for technical snow-jobs. The basic cause for disagreement between the expert and the writer lies in their personal objectives. The expert desires to impress people with the value of his work, and his own technical skill. Regardless of how he may try to otherwise convince everyone, including himself, deep inside he regards the article as an instrument to build up his own reputation and status as an advanced technical man. The writer, on the other hand, has little personal axe to grind. His goal is to explain the subject matter as clearly as possible. He wants to make it appear simple, and in this aim he meets another barrier, because when the subject is made simple, the status of its inventor is also reduced.

In any conflict that may arise in the collaboration between expert and writer, one fact is painfully true: The article is really the expert's; the writer is only a ghost, a consultant. If he cannot convince the expert of a given wording, he must bow to the expert's will, no matter how wrong it is. The writer must never become emotionally involved in the job. By preserving a detached attitude he will have the most influence on his tempermental collaborator. Like the staff assistant to a general, he must rely on gentle persuasion to accomplish his goals.

We have implied that writing for a technical journal involves material difficult for the average technical writer to understand. This may be true of journals dealing with science, but it is not so in the case of many journals. Magazines in the technical fields of materials handling, safety, and production technology are easy for any competent technical writer to write for. In many cases the writer needs only get the expert's material in a report and a few conversations, in order to organize and write an excellent article. Since some journals pay small fees for articles, a writer can supplement his income with short pieces about production shortcuts and ingenious ways of doing·things which he casually observes around him.

Articles written for the popular journals present a challenge entirely different from those prepared for the sophisticated magazines. The challenge is not in the accuracy of presentation of difficult ideas, but is in the clear and entertaining explanation of simple things. The writer must stay within the vocabulary and general knowledge of the technician, the high school or college student, and the basement hobbyist. Here's an example from Popular Electronics (May, 1962):

> The Signal Monitor covers the frequencies from 80 to 10 meters (including the Citizens' Band). When operating as a field strength meter, it provides a visual indication of the transmitter's RF output. When it's set for audio monitoring, you can check the quality of your AM or CW signal using either a built-in speaker or a pair of headphones....

In this piece the writer assumes that the reader knows what RF, AM and CW mean. Strictly speaking, he should have said "wavelengths" instead of "frequencies" since he mentions meters rather than megahertz. However, amateurs tend to prefer to use the one word, and so an error like this is not serious. The informal style using "you" and "it's" gives the reader a sense of personal relationship with the writer; such style would not be used in an article written for a magazine like Physical Review.

Free-lance articles accepted by popular technical journals are more likely to be written by technicians with writing ability than by professional writers. A technician studying or interested in technical writing may find that writing pieces for popular journals can bring in a supplemental income and increase both reputation and writing ability.

The technician journals are a special kind of technical magazine. They do not pretend to be as scientific as the professional journals, but they still maintain a professional standard in contrast to the hobbyist magazines. The technician journals are read seriously by technicians who desire to increase their competence. They seek to be as accurate and authoritative at their level as the most scholarly scientific publications are in theirs. The articles are not written to be entertaining, but to instruct. They are serious in approach and style. In writing for a technician journal, the writer should bear the following considerations in mind:

1. Make the article a smooth narrative, developing each idea logically from what went before, so as to lead the reader easily to the various points.

2. Use a vocabulary suited to technicians. Avoid unnecessary esoteric technical terms and when in doubt define unusual words, either in footnotes or in the context.

3. Keep the article interesting. The professional reader of an article in a fully professional journal may be willing to wade through difficult and unclear writing because of the importance of the subject to him. The technician is apt to be less determined. If the reading becomes dull, unclear, or difficult to follow, he may decide that some other article is just as valuable to him.

It is hardly necessary to note that manuscripts submitted to technical journals should follow the same standards of format that apply to other types of professional manuscripts. They should be double spaced and typed on one side of the pages. The name and address of the author should be placed on the first page, generally in the upper right corner. Diagrams and linework illustrations should be submitted in india ink on white bond stock, suitable for photo reproduction, and photographs should be supplied as glossy prints, preferably in 8x10 size. There are numerous texts and manuals on manuscript style. In a later lesson (Technical Editing) we will discuss in detail the format of a technical manuscript.

Mass-Consumption Articles on Technical Subjects

In the last few decades, science and technology have emerged from specialized **anonymity**.   Articles on technical subjects are appearing with increasing frequency in the mass-circulation general magazines, and are accepted by the public as readily as articles on politics or sex.

Three-quarters of a century ago, a "science" article in a general-circulation magazine was likely a lurid description of various ways in which the end of the world might occur. The scientist was often visualized as a madman who created monsters or worked out schemes to go to the moon, and the engineer was the man who ran the steam locomotive. Today science has caught up with the imaginary feats of a few decades ago. There are still articles about trips to the moon, but now these are factual news stories.

Articles on technical subjects now appear in such general magazines as Readers Digest, Life,* and the news magazines like Time and U. S. News and World Report. The magazine sections of the Sunday papers, long the purveyors of lurid science, now publish articles that were once found only in Popular Mechanics. Sophisticated magazines like Esquire, which are aimed at highly intelligent readers, often run articles of a technical level equal to that of a trade journal in the field (for example, on high fidelity audio equipment).

Science articles in popular magazines have grown with science itself. In the nineteenth century, scientific articles were concerned with polar exploration and the transAtlantic cable. Today the general reader has shifted his interest to the exploration of the moon, and the latest planetary probes. The emphasis on the scientist as a personal hero has declined because science itself has grown to an immense team effort in which the exploits of any one man are small in contrast to the overall accomplishments. Even the excitement created by the first heart transplant has broadened until the original hero has blended into the many medical scientists who made the wonderful technique possible.

*This lesson was written in 1970. At that time, *LIFE* was a prominent magazine with extensive U.S circulation. This footnote was added in 1976, when selected lessons in Technical Communication (from the Grantham Engineering Series) were reprinted in this book.

A characteristic of present day science writing in contrast to that of a few decades ago is that it is more responsible. A quarter century ago, unproven hypotheses were blown up to epochal proportions, and the public looked on the scientist as a kind of mountebank who had inherited the cloak of the medieval sorcerer and alchemist. Today the writer of popular science pieces assumes the same kind of responsibility for accuracy that the scientist himself feels, with the result that science writing and reporting enjoys a much better reputation. In 1910 high school students hid science articles along with their dime novels. Today an article in <u>Life</u> on geology or anthropology will be used by high school teachers as supplemental material in their science classes. The new respect for science is due in part to the responsible attitude taken by science writers.

We have noted that the reader of a technicians journal reads an article for the information it contains, rather than for entertainment. The attitude of the general reader is also changing in this direction. He no longer wants to be thrilled by speculation about life on Mars, but to be informed of the experiments designed to seek out such life. This does not mean that general science articles must be pedantic or dull, but only that they must combine entertainment with solid information. The science writer must develop his line of reasoning as a mystery writer plants his clues, so that the reader is as anxious to learn of the conclusion as he is to discover the murderer.

The modern writer of science articles cannot rely on extravagant statements or unsupported opinion, but must give authority and proof for what he says. He may not use footnotes or present a bibliography of references, but he will be wise to have these to back up what he says. Editors are very sensitive about the adverse criticism brought on by factual errors, and a writer is well advised to be ready to back up everything he says with proof.

In writing an article for any type of magazine, the writer must form a mental image of his reader, and then use a style and vocabulary appropriate to this idealized person.

The <u>Scientific Amercian</u> reader is generally a college graduate, employed in a technical capacity, who enjoys learning about advances in fields other than his own. The biologist likes to know about computers, the astronomer enjoys reading about brain surgery, and everyone is fascinated by the wonderful game called mathematics.

The <u>Popular Science</u> reader is generally a high school graduate who may work as a technician but probably is just a hobbyist in technical matters. He enjoys knowing about science in a more superficial way than the <u>Scientific American</u> reader. He wants to know results rather than reasons -- how things work instead of why.

The <u>Life</u> magazine reader is so varied as to make categorizing difficult. The editors may take a dual approach, using simple captions for the pictures, and then presenting an in-depth article on the back pages for those readers who want to know more about the matter.

The <u>Business Week</u> reader is an executive who wants to keep informed on things which may relate to his business. He is intelligent and accustomed to absorbing complex ideas so long as they do not demand specialized technical knowledge. This magazine sometimes offers technical descriptions which appear, at first glance, to be intended for experienced technicians rather than intelligent laymen. For example, consider the following which appeared in February, 1961:

> The telephone system of the late 1960's and 1970's may well come out of ESSEX, which stands for Experimental Solid State Exchange... Pulse Code ESSEX, just now going from research to developmental status, is really revolutionary. It transmits everything -- switching, signals, voice and data -- in the form of a pulsed code. By sampling voice transmissions 8,000 times per second, it can send out coded bits of information that can be recreated into a voice signal equal to the average telephone hookup... In effect, it does electronically what you do when you draw a smooth curve through a series of dots plotted on a chart. From the individual bits of information, it smooths out pulses and yields a smooth voice tone. Scientists call this "pulse code modulation".

In the foregoing example, the only technical knowledge required is that of smoothing a curve by drawing lines between plotted points. Such curve smoothing is done by technicians, but it is also well-known to business planners and is therefore a good comparison to make in explaining the sampling technique of pulse modulation. Words like "sample" which are used both by electronics engineers and businessmen are useful in bridging between the two fields. The new technical phrase "pulse code modulation" which is unfamiliar to the average businessman, is carefully explained. The reader of this article comes away not only informed of a new development which may affect him, but given a bit of technical jargon which will help him in talking later with his engineers.

Study of the foregoing example also shows a writing style that is very effective for material of this nature. The sentences are short, and each presents one basic idea, with emphasis to make the idea stick. Dashes are used at one point to present footnote-type material in a rapid-fire manner. Writing techniques like this are psychological tricks to enliven the material.

The use of short sentences is good in popular science articles. When unfamiliar words are used, a short sentence presenting only one idea makes comprehension easier. Another technique used by the experienced writer is to avoid pronouns which refer to an unfamiliar technical term. If the pronoun "it" refers to a noun-with-modifiers like "digital sampling technique" earlier in a sentence, the reader may become so confused that he has to backtrack to the noun to get his bearings. It is better to use the noun twice in two short sentences, rather than to use it with a pronoun in a longer sentence.

The use of short sentences can be overdone. Too many of these make choppy reading. They also make dull reading. Too brief a sentence can't develop a complex idea. The mind has trouble integrating many brief ideas. In the extreme, short sentences become like a child's primer: "This is a transistor. It has three leads. One is the emitter. One is the collector. One is the base."

Fairly long and involved sentences are used in these lessons. In the preceding paragraph, short sentences were deliberately substituted. The result, we feel, is somewhat choppy. A longer sentence with several dependent clauses, requiring a semicolon or so, gives a feeling of leisure and smoothness in writing; the reader can pause and reflect before going on to the next sentence, and the overall result is at least more elegant than the truncated stacatto style affected by some writers when they write for an audience of morons with short attention spans and shorter recall. (We overdid it a bit on this one, tho!)

The science writer is constantly confronted by the dilemma of entertainment vs instruction. In the past, science pieces had to entertain first, and writers evolved the showy sensational style of the 1920's and 30's. Today entertainment is still needed, but the reader is now more sophisticated and doesn't demand a miracle in every paragraph. The development of a technical vocabulary, in the average reader, however, is still slow. A sports writer can use very specialized terms to describe a situation in baseball and football, and expect that all readers interested in sports will understand them. A science writer has to be more careful. If he throws in even a few unfamiliar words, he will lose his audience. If you try writing a technical article for a general circulation magazine, you must choose your terms carefully, and know when to omit or when to explain.

Planning Popular Science Articles

The most important aspect of writing about science or technology for the general reader is concerned with the overall design of the article. In order to plan and write at all, the writer must know a great deal more about the subject than he will include in the article. Such knowledge gives room for selection in methods of presentation and examples, and lends a subtle air of authenticity to the whole presentation. If the writer puts everything he knows into the article, it will degenerate into a string of facts presented one after the other. If he backs his presentation with a broad knowledge of the subject, he can select and employ facts to build up a particular concept.

A science article has definite objectives. These are to create an understanding of certain technical matters which lie within a larger field. One goal of the article is to relate these details to the total field -- that is, to set them in proper perspective. Clearly, the writer must comprehend the nature of the total field in order to set forth the relationship between his subject and the overall field. In creating a clear understanding of a large subject, a writer must put together many small elements of fact, like a bricklayer putting up a wall. When broad generalities are stated to a reader unfamiliar with the supporting details, a feeling of vagueness and frustration is generated, and the reader is apt to end up doubting that the writer even knows what he's talking about.

The need for relating a detailed discussion to a larger field is more important in articles for general magazines than in articles prepared for specialized journals. In the latter, the reader is presumed to know the relation, and to be able to generalize from particulars. In general magazines, the first question in the reader's mind is, "How does this affect my life?" If the writer shows how his subject affects society in general, then he will have answered the personal question for most of his readers.

A writer can get technical details from specialists whom he may know, or from textbooks. He cannot even start compiling information, however, unless he is aware of the totality of which his subject is a part. Without such general understanding, he can't even ask the right questions of the specialist. For example, if a writer wants to describe the synchrotron in an article, he must understand at the outset what this machine is for, how it works in general, and specifically how and why it differs from the cyclotron. These, of course, are the key points to be explained in the article. The steps by which the writer builds up answers to the large questions in his reader's mind are also the questions which he must get answered by experts or discover in reading.

The best way for one to learn how to write popular science articles is to study successful examples. To do this, read a chosen article in a magazine like Life, and note the points discussed, perhaps using 4x5 cards. Put the cards away for a week or so -- giving enough time for you to forget exactly how the article was prepared. Then, referring to the cards for facts, write your own article, and compare it with the original. If your memory is too good, ask someone else to read an article (which you haven't read) and prepare notes on cards, from which you can write your own article for comparison.

A good science writer is a teacher with thousands of pupils in the persons of his readers. Just as the effectiveness of classroom teaching is increased by personal enthusiasm on the part of the teacher, so the impact of an article is greater when the writer approaches his task with zeal and personal involvement. There must be created a sense of excitement --even urgency -- in the reader, if he is to have his interest captured.

Every teacher knows that points are clarified best when they are related to things the students know or are interested in. Here are two ways to explain Bernoulli's principle in fluid mechanics.

1. Bernoulli's principle, discovered in the seventeenth century by Daniel Bernoulli, a Swiss mathematician, states that the pressure of air moving over a surface is reduced in proportion to the square of the velocity. This principle is basic to many physical phenomena, such as the lift generated in airfoils....
2. How does an airplane fly? Why does a baseball curve when it is thrown the right way by the pitcher? These effects, and a lot of other phenomena that we meet with every day, are explained by a simple physical law called Bernoulli's principle, after the Swiss mathematician who discovered it....

In the second example, the two questions relate the subject to matters of interest to most people, and make the reader want to go on to see what the answers are. The first version starts out by compounding a dry beginning unrelated to the present day, with a confusing statement in mathematical terms. An article about Bernoulli's principle has got to get over the fact that pressure decreases with v^2, and the writer has got to use his imagination to decide on how to do this. Just saying "directly as the velocity squared" is out -- it loses the average reader immediately. How would you do this? Think this over, and jot down your ideas -- or better, write a paragraph as you would in an actual article. Call this exercise Practice Problem #1, and when you have finished, take a look at the end of this lesson for our solution".

The science writer should realize that he is telling his readers about wonderful things. He is in a very real sense, a latter-day prophet, and he must approach his task with romance, a touch of wide-eyed naivete, and a bit of the mystic. While writing, he should create in himself the sense of wonder that he wants to instill into his readers. He must be awed by the vast extent of time and space, amazed at the temerity of scientists in attacking their great problems, and filled with anticipation for the next exciting chapter in the world's greatest adventure serial, the cliff-hanger called science.

If you feel that we have presented so many conditions for good science writing that you cannot possibly qualify, take heart. Few other writers can either. We can recommend very few good science writers, who combine knowledge with writing skill and enthusiasm for the subject. So you're really in the same boat with just about everyone else, and you have just as good a chance as anyone else to become a successful science writer. Science writing is still a wide-open field, without the competition found in sports writing, for example. Writing about science is much more difficult than exposing the sex life of a movie star, but there are fewer writers trying it (and it's really much more interesting to write about science). There is little or no formal training available in science writing, but if you combine moderate skill with moderate knowledge about science and enthusiasm for the idea, then you are in an excellent position to break into popular scientific writing.

So if you are interested, study examples of good writing, look over professional writers' magazines like Writer's Digest which lists the requirements of magazines, and after deciding on which market you want to crash study the articles published in it. Then go to it, and don't get discouraged at a few rejection slips.

Newspaper Science Stories

Newspapers publish stories about events in the world around us. These events include happenings in politics, industry, the personal lives of famous people, and above all, violence and crime. The impact of science and technology on life has made the science story important in newspapers. A science story may be a reflective piece, like a magazine article, but more often it is a report about some new and dramatic discovery or invention, which is handled on the same basis as any other report of current interest.

The writing of science stories is complicated by the fact that the writer cannot assume background knowledge on the part of his readers. When a crime is reported, for example, the reader is expected to know in general the situation producing crimes; there is no need for the reporter to outline in detail the social climate of the area in which the crime was committed, or the political situation which may have had an effect. In science, people do not have a common knowledge of such backgrounds, and even terms must be defined in the context of a story. All of this must also be done without slowing the pace of the story, or taking too much space. Such writing presents a real challenge to the ingenuity of the writer in thinking up analogies and fresh dramatic ways of expression.

Most large newpapers employ a fulltime science editor, who prepares most of the science stories. Smaller papers usually assign a science story to some reporter who has a technical background or knows people who can give him background information. The smaller papers are also interested in making a relationship with a free-lance writer (called a stringer) who can be called upon to prepare a story on a specialized subject, or who will keep his eye open for newsworthy events which he will write up and submit.

Many newspapers run regular science columns, either written locally or purchesed from a syndicate service. These are generally more like articles than stories of current events, since they are written weeks or months in advance of publication.

Syndication offers an interesting opportunity to the free-lance writer. If he feels that he can write a story each week on a long-term basis, he should prepare a backlog of, say, 25 stories, and submit these to one of the national feature syndicates. If accepted, the syndicate will undertake to sell the stories to its newspaper customers. The price varies depending on the circulation of each paper buying the column, and is generally low -- say, under five dollars per story for medium sized papers. This rate of pay may appear very low, but if a score or more papers buy the same material, the profit for the writer can become very attractive.

Another intriguing fact about syndication is that if the big syndicates turn down your column, you can syndicate the material yourself. The mechanics are very simple. You print up, by mimeograph or offset or even spirit duplicator, 50 or 100 copies of each story, and send them out to prospective newspaper buyers. For example, send five stories with a letter saying that they can try these samples free as a test of reader response. If they want to go on, you will then supply them with a story per day, or per week, for as long as they wish. If only two or three papers buy, you may have quite a lot of work for little return, but then all businesses start slowly -- even the big syndicates were once home operations -- and if your work is good and you are persistent, there is no reason why you too should not have a share of the profits of the great boom of interest in science.

When a science story breaks, and you have the job of reporting it, how do you go about assembling information? Like any reporter, you seek to interview people who know, perferably those directly involved in the story. If the latter are not locally available, go and interview local professors at a college or technical people in industry who might know. Read up on the background (encyclopedias are fine sources for this) and see what you can find in some of the professional journals in the field of the discovery.

If you are going to write about science, you should educate yourself in this field. Read books on science, subscribe to magazines like Scientific American, take evening courses in science in your local school or college, and take correspondence courses. Build up a background of information and interest, and you will find that ideas and writing jobs will naturally start to come your way.

Practice Exercises

The best practice in writing is to write. Earlier in this lesson, you were given one fragment of science writing for a slight test of your ability to formulate scientific ideas in a form that the layman can understand. The following practice exercises give quite a bit more of a workout, both in writing and in the gathering and organizing of material. The examples given at the end of this lesson are taken from a group of more than 100 short popular articles called Science Capsule, prepared by the writer of this lesson for newspaper syndication. Most of these were also taped for broadcast under the auspices of the University of Washington.

Each subject should be written up in 300 to 500 words, in a style aimed at the average newspaper reader. There should be no complex technical discussions, no mathematics, and no assumption of knowledge beyond what the average man might know. The style should be light and informal, with even a little humor (but not too much). Most of the subjects deal with concepts that you, as an engineer, are at least familiar with. In preparing these little stories you should draw first upon your own knowledge and experience, and then look up information in textbooks, encyclopedias, and other reference sources. If you know a science teacher knowledgable in the field of one of the stories, try out your interviewing technique. Here are the subjects:

2. Describe the concept of absolute zero. Why can there be no lower temperature? Who first discovered absolute zero, and how did he reason it out? On what temperature scale is zero equivalent to "absolute zero"?

3. Tell in 300 words how the atom bomb works.

4. The actual brake horsepower of an automobile can be measured by use of a road test and a simple formula. To make the test, first weight the car. Then run it at a speed 5 mph above the speed at which you want to know the power. Put the car into neutral and allow it to coast until the speed is 5 mph below the required speed. Then horsepower is $HP = 1.21(MV/T)$, where M is the weight of the car in kilopounds (e. g. , M = 3 for a 3,000 pound car), V is the desired in mph, and T is the time in seconds for the coasting to occur from $V + 5$ to $V - 5$. In your writeup do not give a formula, and of course, avoid any extended discussion of theory. Just tell how to make the test and how to do the arithmetic. (Incidentally, you might enjoy working out the derivation of the formula

itself. It's based on Newton's laws of motion, and is not difficult at all to do.)

5. Tell some facts about dry ice -- what it is, how it is made, and what its uses are in industry.

6. Describe the Heaviside layer, as to what it is and what it does.

7. Write a story about screws -- machine screws, wood screws, power transmitting screws -- and how important they are in technology and daily life.

Solutions to Practice Exercises

In the subject of our lesson, we can't really offer "solutions" to the practice problems or exercises. There are a thousand ways of working out a problem in expository composition, and you may come up with something much better than ours. However, here are some ideas:

1. Bernoulli's Principle

The best course here is to build up the idea of variation according to the square, by giving some numerical examples. Try this approach:

If the air moves at one foot per second over an area of a square foot, the pressure is decreased by about a thousandth (0.001) of a pound. If the air moves at 10 feet per second, the change in pressure is not just 10 times greater, but 10 x 10 or 100 times greater, for about 0.1 pound per square foot. If the velocity is hurricane speed, say 100 feet per second, the change in pressure is 100 x 100 times as much, which is 10,000 times the 0.001, or 10 pounds per square foot. If this kind of breeze is flowing over a house roof having a surface 40x40 feet (1600 square feet), the drop in pressure will be 16,000 pounds, which is enough to lift the roof right up into the air!

2. Absolute Zero

We've all heard of a temperature called absolute zero, and those of us who remember our high school science may recall that this is supposed to be 459.72 degrees below the ordinary Fahrenheit-thermometer zero. Such temperatures are never encountered on earth (−93 in Siberia is the record), yet this lowest possible temperature was known quite accurately long before it was attained in the laboratory. How this was done provides a good example of scientific detective work.

Back in the 1600's an Englishman named Robert Boyle discovered that if a container of fixed volume containing air was cooled, the pressure of the air decreased. Translated into modern units, he observed that when the vessel was taken from the warm laboratory at 87 degrees, outdoors where the temperature was 32 degrees, the pressure dropped by 1.4 pounds per square inch. Knowing that the original pressure was 14 pounds per square inch, he calculated that the air had lost one tenth of its pressure when the temperature fell 55 degrees, from

87 to 32. If things kept on this way, a total temperature d r o p of 550 degrees should reduce the pressure to zero and the air would be gone! A bit of subtraction told Mr. Boyle that 87 minus 550 gave 463 degrees below zero. Since a negative air pressure is a bit illogical, this temperature was given the n a m e, Absolute Zero. Later and more-accurate experiments have served to verify and strengthen this basic concept arrived at so simply in its original form.

Absolute zero has never been reached in the laboratory, but scientists l i k e Prof. Paul Higgs at the University of Washington have gotten to closer than 0.01 degree of it, which in this writer's opinion is too close for comfort.

3. How the Atom Bomb Works

Sooner or later any science column must get to the atom bomb, so h e r e we go: nuclear fission in 300 words!

All atoms are made up of nuclei surrounded by electrons. The nucleus is composed of several kinds of particles some o f which carry positive charges. Positive charges repel each other, and the nucleus would fly apart violently were it not for other forces of attraction within the nucleus. The natures of t h e s e forces are not entirely understood by physicists. The situation is somewhat like a set mousetrap which would like to snap but cannot because of a locking device. All the mousetrap requires to release the energy in the spring is a mouse. Similarly, all an atomic nucleus needs to blow up is the right kind of stimulus.

A few atoms are so complex and overcrowded inside that they tend to fall a-part spontaneously. Radium is a good example of this. Every so often there is some kind of internal fight and a radium nucleus breaks a p a r t i n t o several smaller nuclei which go flying off with considerable energy. Uranium does this also, to a lesser degree, and one kind of Uranium -- the kind called 235 -- goes to pieces everytime it is tickled by a kind of atomic particle called a neutron. If we had enough neutrons, we could spray Uranium with them a n d really cause havoc. The problem is to get enough neutrons. A t t h i s point Einstein, Fermi and Meitner has an idea. There are some neutrons in the Uranium nucleus which are shot out when it breaks down. Why not arrange things so these could tickle more atoms and thus get a kind of snow-balling effect with plenty of neutrons for everybody? The idea was tried out (after considerable calculation) in New Mexico in 1945 with rather spectacular results, and the Atomic Age was born.

4. What Horsepower Has Your Car?

In this column we try to avoid that nasty word, mathematics, b u t here is a place where we think a formula is OK, because it makes it easy for car owners to find out exactly how much horsepower it takes to run the old buggy. The answer may come as a shock to some, particularly those with new models boasting 200-plus horsepower. The cruel truth is that, for steady going on a level road, 20 or 25 horsepower will take the average car along at 50 miles per hour w i t h no trouble at all. Only during hill climbing, or momentarily in quick starts, i s even 50 or 80 HP really needed.

To determine the power of your car, say, at 50 miles per hour, you get out on a straight level road and bring the speed up to about 57 or 58 mph. Have someone along with a stopwatch, or at least a watch with a sweep second hand. Now, depress the clutch (or shift your automatic transmission to N). The car will begin to slow down. When the speedometer reads 55, start counting time and continue until the speed has dropped to 45. Record this time in seconds -- for example, it might be 12 seconds.

Next find the weight of the car, as of the time of the test. The best way is to put it on a public scales; otherwise, call the dealer and add to his figures the weight of passengers and gas in the tank at six pounds per gallon. To find the horsepower at 50 mph, we multiply the weight of the car in thousands of pounds by the speed, divide the result by the time, and multiply the whole thing by 1.21. Thus, if weight is 3,850 lbs and the time is 12 seconds, HP = $(1.21 \times 3.85 \times 50) \div 12$ which works out to be 19.4 horsepower. This is what engineers call the true or "brake" horsepower actually delivered at the clutch, and takes account of tire friction, air friction, and any grade. It is as exact as the figures used to find it.

5. Dry Ice

Dry ice is a collection of inconsistencies. It's not ice, it's a solid gas, and it's so cold that it burns. Dry ice is not found in nature, not on earth at least, because it exists only at temperatures lower than 109 degrees below zero. It is an exclusively artificial product made by refrigeration machinery.

At ordinary temperatures the material we call dry ice is a gas. Its chemical name is carbon dioxide, it is produced by flames and as a byproduct of animal life. With every breath we expel some of this transparent, odorless gas into the atmosphere. The air might soon be full of it were it not for the fact that plants consume carbon dioxide -- and in return, give out the oxygen necessary for animal life! Plants thus supply us not only food but the very air we breathe.

Dry ice doesn't melt, but changes directly from a solid to a gas, a process called sublimation by the chemists. At high pressures, however, it will liquify if the temperature is around seventy degrees below zero.

Carbon dioxide is a heavy gas, and flows almost like a liquid, displacing the air. This makes it useful as a fire extinguisher, for it pushes the oxygen out of the way and literally smothers the flames. It is very soluble in water, especially at high pressures. If the pressure on water containing dissolved carbon dioxide is released, the excess gas comes out in bubbles, and we have "soda water".

At the temperature of dry ice, rubber becomes brittle and lead makes a fine spring. It is used to freeze biological specimens prior to slicing them into thin sections for microscopic study.

The gas used for dry ice is obtained from volcanic vents and, oddly, as a byproduct in the fermenting of molasses to make Jamaica rum!

6. Cosmic Mirror

There's a mirror up in the sky which plays a mighty important part in modern life. We don't see it, because it doesn't reflect light -- it reflects radio waves. It is called the Heaviside layer, after the British mathematician who predicted its existence long before radio was invented, and it is neither silver nor even solid. It is a region 50 miles or more above the earth where the air molecules have been beaten up by a cosmic goon squad -- or, more elegantly, by high speed electrons shot out from the sun. In the course of the rough treatment the air molecules lose electrons, which leaves them with positive charges. Molecules so **embarrassed** are called ions, and so this region at the top of the atmosphere is called the ionosphere.

It's not difficult to see why a layer of ions acts like a mirror. Radio waves are just rapidly alternating electric fields, and when they hit an ion they make it wiggle. Now any time a body with an electric charge on it wiggles, it sends out radio waves. This happens when electrons vibrate up and down the antenna of a radio transmitter, or when a charged molecule 50 miles up is made to vibrate. The effect of all the wiggles in the ionosphere is just the same as though there were another transmitter up there, with a thousand mile range because of its height. So we can look at the Heaviside layer as a kind of supplementary radio relay station in the sky or a great mirror for radio. Either way the effect is the same -- the extension of the range of a transmitter by thousands of miles.

The short waves used for TV and FM broadcasting vibrate so fast that the ions can't keep up, and so they won't re-radiate these wavelengths. That is why such signals are limited in range to the horizon, or to the line of sight from the transmitting antenna.

7. Screwy Business

Much has been written about the great inventions which have transformed the world. The steam engine, radio, and the motion picture are cited for having changed our ways of life. Yet there are other inventions, generally forgotten when the applause is turned on, which are equally important. One of these is a simple and humble device, the screw.

Screws are so widely used, so much a part of everything we have, that we tend to take them for granted. But, just for a moment, let's imagine that every every woodscrew, every machine screw, every bolt and stud, should suddenly vanish. What would happen? Well, quite literally, things would fall apart. The typewriter on which this is written would be in a hundred pieces, your automobile would collapse into a heap of junk -- in short our modern world, from electric shavers to the generators at Hoover Dam, would fly to pieces.

It is interesting to check back to the invention of the screw, and see how people did things before these little devices were available. History books usually credit Archimedes with the first demonstration of the helical screw, but like many other theoretical inventions, its potentialities weren't realized until someone devised a cheap practical way of making it. The engineering genius who made machine screws available to industry was an Englishman named Henry Maudslay, and he did it shortly after 1800 by inventing the modern screw-cutting lathe with a power-driven cutting tool. Before Maudslay's time, nails, wedges, rivets, and leather thongs were used to hold things together, because the only way to make a screw thread was to mark the helix on a steel rod (by wrapping a string around it) and then file the thread out by hand. Even at the pay in those days, screws were expensive items. So let's add Henry Maudslay's name to the roster of inventors whose genius has changed the face of the world.

Summary

Technical articles represent a type of technical writing that offers great freedom in style to the technical writer who may feel constrained by the formalism of reports and proposals. Articles may be written for a wide variety of readers, ranging from research scientists to basement hobbyists. The vocabulary, style of writing, and general approach in an article depends on its intended audience, and the planning of articles to reach these readers represents a major challenge to the writer.

We have mentioned the various kinds of publications that accept technical articles: professional journals and technicians journals for the serious reader, popular science magazines for the hobbyist and citizen who wants to know about science for his pleasure, and finally the mass magazines and newspapers which want science material because of the importance of science and technology in daily life. The selection of subjects and the organization and writing styles required in the foregoing kinds of publication vary greatly, and the writer must use his or her ingenuity in preparing an article which will compete with the work of professionals in the general-article field.

Technical articles may be written as part of a writer's work assignment in a company, or they may be freelanced as a supplementary source of income. A few writers make comfortable livings writing technical articles, but in general the pay is not sufficient to make this kind of work more than a casual supplement to a more secure income. The writer who tries his hand at this form of writing, however, may be rewarded by an improvement in his overall professional skill, by reputation beyond what he can obtain as a captive writer, and by deep personal satisfaction.

TEST TC-6

1. Writing articles which are published in recognized magazines usually does very little for the writer's reputation.................................... _____

2. Technical articles are subject to stringent rules as to style and format........ _____

3. Articles on very advanced or complex subjects are usually written by the experts who have developed the material.................................... _____

4. Throughout the world, there are about 60,000 technical articles published annually.. _____

5. The most technically sophisticated articles are published in technicians' journals... _____

6. The technical writer should be very careful in using newly coined terms....... _____

7. Articles written for popular magazines need not be too accurate in their technical content... _____

8. A technical article written for a medical journal would probably use medical terms in describing an electronic device which is analogous to the nervous system.. _____

9. The science writer cannot assume that the reader of a popular magazine has a technical vocabulary equivalent to his knowledge of baseball terms.......... _____

10. Syndication for newspapers can be done only by large well-financed companies specially oriented toward such distribution................................ _____

ESSAY QUESTIONS

Write an article (600 words or less) directed to a junior high school age group, on one of the following subjects:

1. Newton's laws of motion.

2. The physical action of an LC oscillator circuit.

3. The action of a 4-cycle internal combustion engine.

4. The operation of radar.

5. How a transistor works.

LESSON TC-7
Writing Technical Proposals

Introduction

A few decades ago, in the days before World War II, a manufacturing company went through a well-established process in introducing a new product. First it carried out a market survey to determine what response to the product by potential customers might be anticipated. If it appeared that there would be sufficient sales, the company next developed the product, in its own facilities and at its own expense. Large companies maintained full-time design divisions, staffed by scientists, mathematicians, engineering designers, industrial designers, production engineers, draftsmen and technicians. Smaller companies, unable to pay the continuous overhead of design divisions, might employ an engineering consultant group, or an industrial designer who had his own staff.

When the design was complete, the company would begin production, and its sales division would prepare an advertising campaign, and arrange for jobbers, retailers, and all of the other links in the chain between producer and consumer.

The foregoing time-honored procedure is still followed by many manufacturers, particularly in the field of consumer goods. A large segment of industry, however, has entered a very different kind of enterprise, in which the steps outlined above are altered and interchanged until the whole business is quite different. Their operation has brought into existence a new sub-division of technical writing, which calls for a new type of writer, the technical proposal writer.

The technical proposal is a document very different from the technical reports and manuals which you have studied in the preceding two lessons. It is different in objectives, in material included, and in style. The conditions under which the technical proposal is written are very different from the conditions under which other technical documents are written. That is, the writer of the technical proposal usually finds himself in an atmosphere of hurry and tension, with company exectutive looking over his shoulder, worrying whether too little or too much has been included.

Proposal writers are often paid more than ordinary technical writers, and in return for this pay they are often called upon to work 10, 12 or 14 hours a day to meet a deadline, often at the center of a huddle of worried executives.

A proposal is part of the sales effort of a company. In style it must combine the enthusiasm of sales literature with carefully reasoned and presented arguments to influence technically sophisticated and hard-headed customers. Unlike the technical report or manual which is backed by detailed information and drawings, the proposal must describe something that has not yet been done. It must convincingly outline production techniques which have not been tried, with unproven design concepts and basic new ideas that are novel.

Technical reports and manuals are a part of the job of manufacture, and their cost of preparation is borne by the customer. Proposals, on the other hand, must be paid for by the company. They are part of the gamble of business. If they result in a contract, their cost is paid for many times over; if not, the time and effort expended represent a dead loss, except perhaps for some general experience which may be of value in a later proposal.

Proposals are very important documents in a company, particularly if the company sells in quantity to large customers. Each proposal is a vital part of an overall sales effort, and may be the determining factor in the decision of a customer to award a contract or to investigate further. Successful proposals -- that is, those which result in contracts -- can determine the profit of a company, and even make the difference between success and bankruptcy.

There are many kinds of proposals, differing in length, detail, and style. They may range from a few pages -- little more than an expanded letter -- to printed volumes costing thousands of dollars to prepare. Proposal writers obtain information and advice from engineers and company management, but they should be able to make constructive contributions themselves. A department manager or engineer who can write good proposals is a jewel beyond price. A writer who can take major responsibility in preparing proposals is likewise of great value, and such ability may go a long way toward moving him to a position in management.

What is a Proposal?

A proposal is a document prepared by a company to sell its services to a potential customer, i.e., to convince the customer that the company is capable of performing a given assignment in research, design, or production. Proposals are important to companies which do work on a contract basis, rather than manufacture proprietary products for sale to customers through wholesalers or directly. Industry which depends on selling by proposal is relatively new; it has become important in the last several decades. In earlier times a manufacturer would decide on a product which he hoped would find customers. He would design and manufature it on speculation, and place it on the market through jobbers, or with the aid of advertising, salesmen, and any other techniques he thought would result in sales. If his designs were good and his estimate of the market accurate, the manufacturer would be rewarded by a profit; otherwise he would lose, and possibly end up broke.

Companies which get business by the proposal route avoid the troubles mentioned above, but they inherit other problems which are just as likely to produce major ulcers. The technical writer who writes proposals finds himself right in the middle of these problems.

A company which obtains business by means of proposals has as its customers other companies and/or agencies of government. The decision to propose doing work for one of these customers is based on knowing of a potenital customer's need, which the company is able to fill. Knowledge of the need may be gained by personal contact, or by the customer issuing formal invitations to bid which spell out what is wanted. When the customer knows of a number of companies interested in bidding, it sends them invitations. Companies so invited enjoy a certain advantage over competitors, in that the customer has knowledge of their past performance and more confidence in their ability than might be the case with a new or unknown company. The lack of an invitation, however, is no bar to success in bidding. Many unsolicited proposals have resulted in contracts, and most such proposals will draw at least a reply inquiring about the company.

An invitation to bid sent out to a number of companies frequently, does not mean equal opportunity for all recipients. The customer is often bound by law or by the terms of a contract of his own with the government to ask for a number of bids, even on a job which he intends to award to a certain company on the basis of experience and proven ability to perform. For example, a customer might solicit bids to build a part which is described in such detail that it matches exactly the product of a particular company. In such invitations, the phrase "or equivalent to" is used in the specifications to meet legal obligations in competitive bidding, but wise competitors will recognize that they have little chance to take the job away from the favored manufacturer who already has experience and tooling ready to go on the item described.

Different Kinds of Technical Proposals

There are many kinds of technical proposals. Here are a few typical examples:

A precision mechanical-engineering company with its own machine shop might submit a proposal to a major aircraft company to design and manufacture a servo hydraulic flight control system.

The aircraft manufacturer may submit a proposal to an airline for a complete supersonic airplane, costing ten million dollars or more. (This proposal would be a five-foot shelf of books!)

An architectural firm might submit a proposal to a city for designing a shopping mall, or an urban renewal project.

A technical college might propose setting up a specialized training program for a company, or for a branch of the Federal Government.

A professor of biochemistry in a university might submit a proposal to a pharmaceutical company for funds to support research which might result in a commercial product of value to the company.

A company in the computer business might submit a proposal to the National Aeronautics and Space Agency for a pulse-code communications system.

In the constantly accelerating pace of the space age, there are thousands of companies and individuals writing proposals. The great companies, like General

Motors or McDonnell-Douglas, direct their efforts to major operating companies like major airlines, or to the biggest customer of all, the U. S. Government. When contracts are awarded by these customers, they are referred to as "prime contracts". The companies receiving prime contracts often choose to subcontract parts of the job to other firms. The negotiation for the sub-contract is between the prime and sub-contractors, but since the final product goes to the original customer, his requirements are just as binding on the sub-contractor as on the prime contractor. When the customer is the U. S. Government, its specifications will involve everybody concerned, however far removed from the original contract. It is plain that a great deal of proposal writing must therefore be in the style of proposals required by the Government.

Style in Proposals

A proposal is written for a very limited audience, often for a single reader. The writer can safely assume that the reader or readers are sophisticated and knowledgeable in the field. Therefore, he can use technical terms freely, and omit much explanation, unless this uniquely pertains to the solution described in the proposal.

Proposals are usually written in formal style, even more formally than the technical reports and manuals which we considered in the preceding two lessons. They are written for busy people whose jobs are to evaluate proposals and select the best. Clarity and brevity are thus important characteristics of proposals. If a proposal is confusing or long-winded, the result may be that it is skimmed over, or perhaps not even read through before being discarded.

A solicited proposal directed to a Governmental agency is usually prepared to meet a "proposal specification". The specification may be a general outline of what must be included, or it may be a printed form in which spaces are provided for required information. In the latter case there is really no "writing" to do; the writer merely compiles the needed information and puts it in the right places.

A typical proposal specification might ask for a description of preliminary research and design, construction of a prototype, development of production tooling and fabrication methods, testing, and maintenance and repair plans, including the stocking of spare parts. If the company submitting the proposal is not already known to the customer, there may be questions to be answered regarding the financial stability of the company, its production tooling, and the qualifications of engineering and management personnel. Sometimes there is also interest in the availability of skilled labor.

In writing an unsolicited proposal, the writer must decide for himself (with company advice and instructions, of course) how to plan and write. If the proposal is for an existing need of the customer, the emphasis must be on the company's ability to fill this need effectively and economically. If the proposal is for a project originated by the proposing company, then the writer must sell the customer on the idea or item as well as on the company's ability to do the job.

While the writer of an unsolicitated proposal can technically use any format and style he wishes, he will be wise to write in a manner familiar to his intended reader, as this will give an impression of professionalism.

We have not mentioned what might be called an "internal proposal" which is written by one department of a company for submission to another department or to management. Such proposals are common in very large companies. For example, the General Motors Research Center might offer its developments or ideas to the Chevrolet Division in the form of a proposal fully as complete and formal as proposals offered by the company to the U. S. Army. In smaller companies, internal proposals can be written informally, and the dividing line between proposal and report tends to become very indistinct.

Preparation of Proposals

Before a proposal can be written, the information required must be compiled, and the general outline decided upon. Two types of information are desired by the reader of a proposal -- (1) information as to what will be done and how, and (2) information on cost. For the first, the writer will go to the design and engineering departments of his company, and for the second, to the fiscal officers of the company.

In writing a solicitated proposal, the writer is in competition with all other companies invited to bid. The temptation is to make the proposal very complete and elaborate, with detailed descriptions and many illustrations, both charts and engineering drawings, and artists renderings. In companies with complete technical publications sections, there is a tendency to go overboard on a proposal. Usually the expense of such an effort is not justified, for the readers of proposals, especially in the Government, are very wise in the ways of proposal writers, and are likely to be unimpressed by pretty pictures and fine prose. The effect of a very elaborate proposal may become negative, if it is so detailed as to confuse the reader. He is not there to buy pictures, but to evaluate a product, and the picture is useful only to the extent that it describes the product clearly and accurately. There is only one criterion for a paragraph, a picture, or a chart; does it create a clearer understanding of the product?

The effect of written material in building up a clear concept is heightened by good typographical format. By this is meant the division into paragraphs with spaces between, the use of boldface letters and italics in printed copy, and the use of spacing and underlining in typed copy.

The content and organization of a proposal must be decided upon before any writing can be done. These decisions are usually not made by the writer, but by management and the responsible engineers who expect to follow through with the project if and when a contract is awarded. The writer must know these matters thoroughly and also should be aware of the reasons for them, since these can guide his own decisions on the details of style and the emphasis to be used.

Technical proposals in industries such as electronics are usually prepared

under the direction of the potential future project engineer. The engineer, however, may have as his associate a representative of the sales department, since
experience has shown that many engineers tend to obscure a proposal with too
much technical detail. The writer, subordinate to both of these men, sometimes
finds himself in a difficult position when they disagree as to what should be included and emphasized in the proposal. Often when the clash between the engineering and sales viewpoints results in a stalemate, the writer may be appealed
to as a kind of neutral umpire. In such situations the writer should be very tact-
ful, and seek to bring the two combatants together rather than to side with one
or the other.

Ideally, the engineering and sales representatives on the proposal team
should meet with the writer, and the writer should take a more passive role than
the others. In practice, the engineer may be very busy on a prior assignment
and the sales executive also deeply involved in other matters. When such is the
case, the writer may find that his job is to meet with these people separately,
almost like a mediator who then reconciles their viewpoints until he has an outline that both will happily approve. Such a procedure obviously is not the best
way to put a proposal together, but it is frequently followed. The writer should
not complain, but should consider that he is being treated with special confidence
and trust, and try to make the best of the situation. When the writer accepts this
challenge and comes through with a good proposal, he is establishing a very good
reputation for himself.

Getting technical information for a proposal is more difficult than the equivalent job for a report or manual, because there are no design notebooks, models,
calculations or drawings to work from. The material the writer needs is usually
in the engineer's head, and it is the writer's first job to extract it. In the process of extraction the writer usually finds that the engineer has solutions in
mind. A proposal is a better selling document, however, if it first presents the
problem to be solved. The reason is that if the customer can first be sufficiently impressed with the importance and difficulty of a problem, he will better appreciate the **elegance** of the solution. Thus the wise writer will take notes as the
engineer talks, and by questions, gradually trace the path backward to the problem. After this detective work, the writer then sets himself to the job of writing
itself, and builds the proposal outline in the reverse direction -- that is, from
problem to solution.

After a first draft is written, it is submitted to the engineer. He immediately finds all kinds of errors and omissions, and is galvanized into action to make
corrections. This procedure does not trouble the experienced writer, because
his whole goal has been to get the engineer going. His psychology is a little like
that of the teacher who deliberately makes mistakes to stimulate his class into
showing him up.

When the engineer prepares his first draft, the writer usually must edit this
material for spelling, grammatical construction, and general clarity of thought.
The writer may also find technical errors. Correcting these may call for tact on
the writer's part; tack is one of the major factors in the success of a proposal

writer, just as it is in almost any career involving interpersonal relationships. If you find that you have to work with an engineer who is crusty and uncoopera- tive and who tends to ride you, don't respond similarly. Just regard the situation as a professional challenge to you as an expert in diplomacy. A "mean disposi- tion" can usually be traced back to troubles at home, or some other condition unrelated to the job or to relations with you as a writer. If you can break through the armored front and get your opponent to talk to you about his trou- bles, you will instantly ease the situation as far as your position is concerned, and probably have the satisfaction of helping the other person too.

The foregoing comments may seem far removed from technical writing, but they bear on a larger matter which includes writing. Tact and understanding in your writing job will help you develop the most important of all executive skills -- the ability to get people to work together effectively.

When the required information has been assembled from the engineering and sales departments, the writer must organize the basic outline for the proposal into which this information will be fitted. Proposals usually follow a fairly well standardized outline, as shown below.

1. Letter of Transmittal

2. Frontispiece

3. Table of Contents

4. List of Illustrations

5. Summary

6. Section I, Introduction

7. Section II, Statement of the Problem

8. Section III, Technical Theory Discussion

9. Section IV, Planning and General Organization

10. Section V, Exceptions

11. Section VI, Statement of the Job to be Done

12. Section VII, Background and Qualifications of the Company

    a. Past experience

    b. Key personnel

    c. Plant facilities and labor force

13. Section VIII, Financial or Fiscal Summary

14. Appendices

15. Alphabetical Index (not always required)

The Letter of Transmittal is usually not the writer's responsibility. It is prepared in the executive offices for the signature of the company president or other official. It is a formality required by custom, and usually starts with "We are pleased to submit herewith our proposal in response to your invitation to bid...". The writer's job is to see to it that the letter is inserted at the proper point in the proposal.

The Frontispiece is also a formality, derived from the illustrations used in books. It gives an appearance of completeness to the proposal, and also serves a practical purpose in giving the reader a quick general impression of the item proposed. In the case of some exotic space devices, the frontispiece is important in giving an overall picture of something very new in concept and appearance. It is usually an artist's rendering of the overall appearance of the item proposed.

The Table of Contents lists everything in the proposal, including the letter of transmittal which precedes it. It includes the fifteen major parts given in our list, and may also list subdivisions of these when desired.

A separate list of illustrations is given for the convenience of the user who may want to refer to any of these in the course of his reading and deliberations.

The Summary is a very important part of the proposal. It is usually the first part read by the official to whom the proposal is submitted. He uses it to gain his first general impression of the proposal. A well-written summary will arouse real interest in the reader so that he will be motivated to study the rest of proposal with care. The Summary is usually written after the rest of the proposal is finished, when the writer has all of the material clearly in mind. The entire Summary should be on one page if possible, to enable the reader to survey the essentials of the entire proposal at a glance. It should emphasize the most important features of the proposed device or system, and point out the major advantages over other similar systems.

In writing the Summary the writer has a check on the thoroughness of his work in preparing the proposal as a whole. If he finds it difficult to decide on a good summary, he will probably question how clearly he has prepared the prosal. He must see the whole picture completely and in detail before he can extract from it a concise summary. If he cannot do this, then the proposal as a whole may have serious defects.

The Introduction presents the basic problem to be solved, or the need to be filled by the proposed project. Its bias is that of sales rather than engineering, since the purpose is to create a desire on the part of the reader for a solution.

Following the outline of the need or problem, the <u>Introduction</u> should present briefly the approach to be taken toward a solution -- i. e. , how the company proposes to attack the problem. Ideas and material for the <u>Introduction</u> are obtained from both the sales and engineering departments, but the writing is done by the technical writer in charge. The <u>Introduction</u> gives in brief form the same information which will covered in more detail in the sections that follow. Like all introductions, whether to books or public speakers, its intent is to provide a starting point in which the reader has been prepared for what is to follow.

The next three numbers of the proposal (Parts 7, 8 and 9) constitute the body of the proposal.

First comes the <u>Statement of the Problem</u>. If the proposal is written in response to an invitation, this statement may essentially copy the requirements outlined in the invitation. The wording is usually changed, however, so that it is an interpretation by the company of what is wanted. For example, if the invitation states that it wants the problem solved by some specified method "or equivalent" the <u>Statement</u> might indicate one or more of the equivalent solutions which the company would like to use. The <u>Statement</u> sets the stage for what is to follow. It creates a point of departure, an area of common agreement on the problem before the solution is outlined. This part of the proposal can be very important. What has gone before is, in a sense, introductory formality, while Parts 10 thru 15 are likewise auxiliary to the main body of the proposal.

Part 8 is the theoretical discussion of the problem and its solution, or what might be termed the technical approach. This section is the real heart of the proposal, in which the company undertakes to sell both its solution to the problem, and its competence to carry the project forward to a successful completion. In writing this section of the proposal the writer must turn to the responsible engineer for the information, and should strongly urge him to write the actual draft. The function of the writer in this section should be primarily to organize, clarify, and edit the basic material as required.

The part of the proposal entitled "Planning and General Organization" shows how the technical approach is implemented to carry out the work itself. The writer explains how the company will apply theory in the actual "reduction to particle" in designing hardware, developing production tooling, building prototypes, laying out a production line, and establishing a program of inspection and quality control. In this section of the proposal the writer should also describe the kind of personnel to be assigned to the project and how many. It should say how they will be selected -- e. g. , from within the company, by hiring from outside, or as consultants. It should give an organization chart for the project, in which lines of authority and responsiblity are shown, together with the relationship to top management in the company -- e. g. , who in the project reports to whom in general management, how often and in what detail. It should spell out what kind of reports will be made to the customer as the project progresses, and give a timetable showing when each phase of the work will be completed.

Item 10 in our list, <u>Exceptions</u>, is included in a proposal when the company

feels that there are sound reasons for changing some of the requirements submitted by the customer. The specifications called out in the invitation to bid are
often drawn up by men whose experience is as users rather than builders, and
sometimes these can add greatly to cost without any real advantage in the performance of the equipment. For example, a demand for a very close tolerance
in a mechanical part can increase cost by a factor of five or ten. If the company
agrees to follow this requirement, its bid must be correspondingly high. In such
a case, under Exceptions there would be a discussion of the situation. In this,
the proposal would point out that greater tolerances can be used without impar-
ing the function of the system, at an estimated reduction in the price.

The writer recalls a case in which the U. S. Air Force set up an electrical
requirement for an antenna which was theoretically impossible to meet. This
requirement was that the standing wave ratio for the antenna (which represents
the impedance looking into its terminals) should not vary more than a given a-
mount as the frequency was changed by 10% from the resonant frequency. Equations found in any textbook on antenna design showed that such a small change in
VSWR could be obtained only by introducing large losses into the antenna so that
its overall radiating efficiency would be very low. When this error was pointed
out, the specs were quickly changed by some very red faced "experts" in the Air
Force!

Item 11, often called the Task Statement, is of great importance because it
binds the company to the performance of certain work in the proposed contract.
When and if a contract is issued, it will spell out with legal clarity the contents
of the Task Statement. If the company accepts the contract, and then finds that
it cannot meet the terms of its own proposal, it will be in very hot water indeed.
There may be a contract cancellation by the customer (especially if it is the
Government) and often considerable penalities.

In the situation with the antenna mentioned previously, the company failed to
realize the booby-trap in the specifications. It undertook to meet the impossible
VSWR requirement, and when the first test items failed, the government assess-
ed penalties which would have bankrupted the company. It became necessary to
hire a middleman with influence in Washington, and to arrange a series of tests
at the company's expense. In these tests it was proven that similar antennas already accepted by the Air Force failed to meet the new requirements. The penalty assessment was dropped, but when the company attempted to recover its
costs in proving the Government wrong, a "great silence" descended upon
Washington.

If a proposal writer had suspected the validity of the demanded requirement
and asked one of the engineers to make a verifying calculation, a great deal of
trouble would have been avoided. This is asking a great deal of the writer, but
there are many cases in which a little natural caution and suspicion can uncover
costly mistakes before they happen. A writer who can do this will not remain a
writer long; unfortunately for the writing profession, he will soon find himself
helping to run the company.

In most cases the Exception taken in a proposal is not intended to uncover an error in the specification. More commonly, it will point out that the company is equipped to do the job better or at less cost if a minor change is made. Perhaps the company has a specialized machine tool, or a manufacturing process not commonly available. When such a fact is brought to the attention of the prospective customer in a proposal, it will be appreciated and usually accepted. In fact, often a company can make a very good impression by displaying unusual know-how in this manner.

The statement of the job to be done should clearly define the information needed from the customer, and exactly what is to be delivered. If some parts are to be customer furnished, this should be made plain. Delivery schedules are usually required, but if delivery could be delayed by a delay on the part of the customer in supplying information or parts, then delivery of the finished product must be made conditional upon getting what is needed from the customer by a specified date.

Such careful spelling-out of terms and conditions may suggest a lack of mutual trust, but experience in business teaches that feelings can be hurt much more by misunderstanding than by binding agreements made in advance. The day of great enterprises founded on a handshake between gentlemen is, alas, over. Warner and Swasey did this in 1880, but they were unusual men, even for that year.

Item 12 in our list covers the qualifications of the company to do the job proposed. A Qualification Summary for a company is like the resume used by professional engineers in applying for a job. This part of the proposal should briefly trace the company's history, giving emphasis to its experience in work similar to that proposed. A qualifications summary is particularly necessary in proposals sent to government agencies, because the evaluating personnel changes constantly. Almost as soon as a personal relationship has been built up with a contracting officer, he will be transferred and a new man who never heard of your company will be assigned! In the case of proposals sent to companies for whom you have done work before, this section of the proposal may be omitted.

In a proposal addressed to a governmental agency, reference should be made to any other contracts of a similar nature which your company had with that government in the past. Successful performance of work can be the best recommendation your company can have.

A list of key personnel to be assigned to the project should be given, with a resume or professional biography for each person. These resumes should vary in detail according to the position of the man. The project engineer or manager should have his background and experience outlined in more detail than designers, draftsmen, production engineers and others. These resumes should emphasize experience in work similar to that proposed. Production workers need not be listed separately (usually they are not even assigned at this stage of the project), but remarks about the general work experience of company production personnel will be helpful.

A description of plant facilities should stress equipment which would be used in fulfilling the contract desired. This might l i s t machine tools according to make and capacity (e. g. , the swing of a lathe) and specialized tools such as vacuum deposition equipment and high precision measuring and inspection instruments. If very specialized facilities are required, such as large environmental chambers (humidity, high temperature, or vacuum), these should be described in detail. When it is important, auxiliary facilities such as a railroad siding should be mentioned.

The Financial Summary for your company is not usually written by the technical writer, but is prepared in the accounting or finance division of the company. In some companies the general statement of financial position i s prepared as a standardized document which is inserted in a l l proposals w h e r e needed. Occasionally the fiscal summary is prepared as a separate document, not given to the technical evaluators of the proposal, but intended for the official i n t h e customer organization who is concerned with financial ability to perform.

Although the writer does not prepare the financial statement, it is usually his responsibility to see that somebody does prepare it in time for inclusion in the proposal, and that printing and other format matters are in keeping with the rest of the proposal.

Appendices are used when there is detailed information which will be desired by an evaluator who is seriously considering the proposal, but which might interfere with continuity in reading the basic document. The purpose of the proposal is to give a clear overall picture of the subject, and the inclusion o f detailed information, including numerical data, can often break t h e continuity o f thought in the mind of the reader.

Appendices may contain numerical d a t a, s u c h as t e s t results supporting statements made in the main body of the proposal. An appendix might give t h e derivation of an unusual equation, or an alternate method of arriving at a conclusion presented in the proposal. If your company intends to subcontract a part of the work and has received proposals from subcontractors, these might properly be included as appendices. If staff members in the company have published articles on subjects related to the proposal, reprints of these will b o t h clarify the proposal and serve to increase confidence in the ability of the company to do the job.

While in the process of assembling information for a proposal, the w r i t e r may be uncertain as to what will go in the main discussion, a n d w h a t should more properly be a part of an appendix. As the proposal approaches final form, there may be a shifting of material back and forth between the body of the document and an appendix. In making final decisions, the writer must judge topics on their merits, with the advice and assistance of technical and management personnel.

Problems of a Proposal Writer

A technical writer who prepares proposals occupies a curious position in his

company. His work is vital to the success of the company, yet he is usually at
a relatively low position on the management totem pole, both in status and in
salary. He must use his own judgment in matters which affect the effectiveness
of the proposal; yet he must submit everything he writes to higher authority
which has absolute veto power over him.

The proposal writer is involved in the tension and uncertainty of management, but he does not have executive power to take action. He shares responsibility but is not given corresponding authority. He is given a deadline for completion of his job, and then encounters delays and excuses from those from
whom he must get necessary information. He may have the unpleasant experience of coming to work on the last day of his writing job to find that management
has decided to make major changes in its presentation or even to drop the whole
matter. When the result of months of work is dumped into the wastebasket, the
fact that the writer's time was paid for does not prevent a letdown.

The writer of a report or manual is describing things which exist. The proposal writer, on the other hand, is trying to keep up with a fluid situation. Right
up to the deadline there will be new items and ideas to put in, and when the manuscript is in the hands of the printer there will still be a feeling that it is incomplete and premature.

If you fear that you will be in danger of getting into a rut as a technical writer, then proposal writing may be your way out. In a large company, you may
be writing about a missile system one week, a new camera the next, and a piece
of complex medical apparatus the week following. There is no time to become
an expert in any of the subjects you write about, and you will always feel uneasy
and fearful that you will commit some awful mistake. As deadlines approach you
will wish for more time to think out the best way to present ideas, and then give
up and write whatever comes to mind, with a "the hell with it" attitude. Yet,
however hectic the pace is, you must seek to stay calm and do a workmanlike
job. You may complain bitterly about lack of information and cooperation from
the high brass, but in the end you have to do the best you can with what they give
you. Perhaps the best attitude is to realize that they also are frustrated and
harried by headaches much worse than yours.

There is much on the positive side in proposal writing, however. The pay is
usually better than that of other technical writers. The contacts with management constitute an excellent schooling if your ambitions are in that direction,
and if you do a good job you are in the best possible position to be noticed when
the boss is looking for a bright young man. Proposal writers may have ulcers,
but they aren't usually bored and they enjoy the status of aristocrats in their
craft.

An Example of Proposal Writing

We have offered a lot of general advice in the art of proposal writing. Writing proposals, like most other jobs, can be learned only by direct contact with
the practical application of theory. In this case, we mean that you must actually

write proposals. We will give you some practice problems in which you can try your hand at writing a short proposal, and in making decisions relating to parts of hypothetical longer proposals. To introduce these problems, we will end our discussion of theory by offering an example of how a writer might receive the information for a proposal, and what his reasoning would be in organizing this information into a presentation that could become a useful sales document.

You are a technical proposal writer in the BC Corporation, which manufactures electro-optical devices. One bright Monday morning you walk unsuspectingly into your office, at peace with the world (save for a slight worry that now that you've finished your last job, maybe they don't need you anymore). As you sit down, the phone rings. It is the secretary for the group of writers of which you are a member. She says: "Mr Becerra has called a meeting for nine thirty. He wants you there."

At the meeting Art Becerra, the chief engineer, introduces Jeff McCoy, vice president for sales; Ed Simmons, who has just completed a project; and six others from sales, engineering, and production. It is quite a collection of brass, and they are gathered to set into motion a major effort by BC. You are not present to enter into any of the decision-making, but your purpose as the designated technical proposal writer is to pick up background information.

Mr Becerra outlines the situation quickly so that everyone will start off with the basic knowledge. It seems that Mr Stephen White, the company president, had lunch last week in the Pentagon with General Faulkner of the Signal Corps. General Faulkner passed the word unofficially that the Army was looking for a compact facsimile system that would transmit photographs over conventional telephone and radio equipment. It had to be small, light, and self-powered, and able to take rough treatment. On his return, White talked to Becerra, and last Monday Art submitted some design ideas. These were kicked around for a week by the management team, and on Friday the decision was made to do some preliminary design and submit a formal unsolicited proposal.

At this point in the discussion Mr Becerra introduces Mr Simmons who has done the preliminary design, and he sketches out how the device will work. After he has used the rather ponderous name "facsimile equipment" several times, McCoy of sales clears his throat a couple of times and suggest that perhaps it is not too early to come up with a name, and how about trying for size, Pictophone? This suggestion sidetracks the technical description and everyone gets into the act. After a dozen versions have been offered, Dallas Inman of sales suggests Telepix, which everybody likes. So from then on, it's Telepix, although Mr Fairchild from the legal department warns that a check had better be made of registered trade-names before going too far.

During all of this conversation you have been making notes. A general picture of the project is forming in your mind. When the general conversation slows down, Art Becerra asks if you have any questions. You ask about one or two points that aren't clear, but you feel very unsure about so many things that when the meeting finally breaks up, you heartily wish that you could get out of

the whole mess. In this opinion you are joined by Ed Simmons, the project en-
gineer-designate, who already sees a number of very sticky problems to be
solved if a contract is granted. This isn't your job, however, so you retire to
your own cubicle, put a piece of paper in the typewriter and tentatively start
putting words on paper. You skip the formal first sections of the future proposal
and start with Item 6 of our list, the Introduction.

As we have noted, the introduction provides a point of departure for the detailed sections of the proposal which follow. Its emphasis is on the problem to
be solved, and the general requirements which must be met. Here is how a first
draft might read:

Section I. Introduction

This proposal describes a lightweight portable facsimile system suitable for
office or field use in the transmission of documents, drawings, maps, and other
line copy. Facsimile systems have been in use for several decades, in both
military and civilian applications, but these are characterized by weight, de-
pendence on external AC power, and the need for electrical connection to wire
lines. The system described in this proposal is light in weight, self-powered by
its own batteries, and is acoustically coupled to ordinary telephone or radio e-
quipment so that no modification or adjustment of the latter is required.

The system here proposed is referred to by the name, Telepix, as a conven-
ient word descriptive of its function. The Telepix system is characterized by
simplicity of operation, with received copy immediately available at the end of
a transmission. It employs a frequency-modulated signal, restricted in range
to the 300 to 3,000 Hz band which is characteristic of conventional telephone
and radio voice circuits. It transmits in a format five inches wide by any length
up to 500 feet.

The Telepix is proposed as a solution to many problems of tactical communication, including the transmission of orders, sketch maps, documentation relating to supplies and personnel matters, and technical drawings and sketches of
equipment. It has a limited capability in transmitting continuous-tone photographs.

At this point you stop your writing on Section I. This brief statement will be
submitted to engineering and sales for comment and probable addition.

Next you commence on Section II, the Statement of the Problem. You have already mentioned this in the Introduction; now you fill in more detail.

Section II. The Problem

The solution to any problem in facsimile transmission must operate within basic limitations set by the electrical characteristics of the associated communication system. The nature of line and terminal equipment in telephone and radio systems is based on the need to restrict the frequency band to the range required for intelligible voice transmission. This band is universally limited to a lower limit of about 300 Hz and an upper limit between 3,000 and 4,000 Hz. Conservative design should take the 3,000 Hz limit.

Voice transmission systems, particularly open-wire lines laid on the ground and military radio, are subject to a variety of distortions and extraneous noise. In voice communication these effects must be at a high level before intelligibility is impaired. In picture transmission, the noise is particularly disturbing, as it appears as spots, patterns, and an overall granular effect referred to as "snow" which obscures the picture, even at a level that would be tolerable in voice transmission.

The solution to the problem of extraneous noise suppression in picture transmission is found in the use of a frequency modulated carrier. Such FM systems are employed in all facsimile equipment.

The use of a FM system extracts a penalty in the available frequency band. The maximum modulating signal that can be used is one half of the FM carrier frequency. Restriction of the latter to a maximum of 3,000 Hz limits the "video" signal to 1,500 Hz. Picture information must be restricted to a band-width of 300 to 1,500 Hz. The scanning format and the picture resolution in the five-inch dimension is thus determined by the relationship:

$$\text{Resolution in lines} = \frac{1,500}{\text{Scan rate in sweeps/second}}$$

This part of Section II is ready for review by engineering; it will be designated as II.1, under the heading of Basic Considerations. The next section clearly should be concerned with the methods used to scan copy for transmission and to reproduce it at the receiving end. These will be parts 2 and 3 of Section II.

II.2 Transmission

In transmission, a scanning point must move across the five-inch dimension of the picture repeatedly while the copy is advanced at a constant rate to provide the scan along the other axis. This rate must make a line separation compatible with the resolution achieved in the five-inch dimension. If a resolution of one "bit" of information is required in a given distance, then two scans must be made -- one to provide the "black" and the other to provide the "white" which are needed to constitute one "bit".

The requirement met in this proposal is the capability of accepting copy in the form of a strip 5 inches wide by any length, or as flat copy of any size, which can be transmitted in sections for paste-up assembly at the receiving end. The flat copy must be transmitted without the need for bending, so that pages of a book, for example, may be transmitted without removing them from the book.

II.3 Reception

Reception must be on untreated paper, supplied in a roll 5 inches wide and in lengths up to 500 feet. The received copy must be immediately usable without processing of any kind, and should be visible during the course of reception.

The advance of the paper is at a constant rate and is equal to that of the transmission of original copy. The mechanical scanning must be corrected at each scan to insure synchronism between the transmitter and receiver.

There must be provided means for stopping the scanning and advancing the receiver automatically when the transmitter completes a transmission. Indicator lamps will show when the receiver is in a "ready" condition. A battery meter will show the condition of charge of the battery used for self-powered operation.

A self-contained charger must be provided for the battery.

Section III outlines the general principles underlying the proposed solution to the problem of meeting the requirements set forth in part II. In the present example, the following brief discussion would be typical:

Section III. Theory of Operation

A. Transmission

The copy to be transmitted will be moved under a scanner consisting of a light source, a lens system forming a point image on the copy, and a photoelectric cell to receive light reflected from the copy. The scanner unit will be moved across the width of the copy (the five inch dimension). A resolution of 500 lines across the five-inch dimension requires a sweep rate of three per second. An equivalent resolution along the copy strip is 100 lines per inch, requiring 200 passes of the scanner per inch. This rate requires that the copy be advanced at a speed of 3/200 inches per second, or one inch in 67 seconds. Transmission time for a four-inch section of copy, corresponding to a 4x5 picture, is 268 seconds, or 4 minutes and 28 seconds.

B. Reception

Copy will be received on a copyset consisting of a "sandwich" composed of 20-pound bond paper and a carbon sheet. The receiver unit scans the copyset in synchronism with the transmitter. The receiver will consist of a diamond stylus attached to an electromagnetic driver, which presses against the carbon sheet to make an impression on the paper.

C. Electronics

The picture signal generated by the photocell in transmission is used to modulate the frequency of a 3 KHz carrier. The resulting FM signal is applied to a low-pass filter which removes the sidebands above 3 KHz, and is then applied to a loudspeaker. The sound from the loudspeaker is picked up by the transmitter on the telephone handset as an audio signal in the range 300 to 3,000 Hz.

In reception, the FM signal sound from the handset receiver is picked up by a microphone and demodulated in a FM detector. The demodulated picture signal is then applied to the electrodynamic drive of the receiving impression unit.

D. Mechanical

The scanner, on which are mounted the lamp-photocell system and the receiver-impression unit, is moved alternately across the five-inch dimension of the paper by means of a reversing motor. The copy strip is advanced by means of a drum and pinch rollers, driven by a second motor.

E. Power Supply

The two drive motors and the electronic components of the system are powered by a 12-volt battery. When the system is operated where AC power is available, a built-in transformer and rectifier produces 12-volt rectified direct current to operate the system and recharge the battery.

Section III does not discuss any electrical or mechanical details of the system, such as the reversing motor mechanism or the type of photocell required. Such details are not properly included in most proposals, since they would be determined in the course of the job proposed. Often such details are known before the proposal is written, but are not revealed in the proposal, since the company hopes to recover costs already incurred during the design phase of the desired contract.

Section IV indicates how the theory of Section III will be reduced to practice. In the present example the following might be included:

Section IV. Planning and Organization

The design-development of the proposed facsimile system will be done in the following phases:

Phase A. Schematic circuit design.

Phase B. Breadboard construction of electronic circuits.

Phase C. Mechanical design

Phase D. Construction of engineering prototypes.
1. Mechanical prototype with breadboard electronics.
2. Mechanical prototype with circuitboard electronics packaged in mockup approximating proposed final external envelope.

Phase E. Design of controls and external case for unit.

Phase F. Detail design of circuitboards and mechanical parts for production.

Phase G. Planning of manufacturing and assembly sequence.

Phase H. Production run of 100 units.

Phase I. Time study of production and firm cost analysis of production.

Since this proposal is unsolicited, there will be no <u>exceptions</u> taken to the requirements of an invitation to bid. Here would be the appropriate place to note that the scanning system to be used is a proprietary item, covered by patents assigned to the BC company. (Such a reservation is needed to protect the company from the practice of the government of farming out production of items designed by one company under a development contract, to other lower bidders when the time comes for large production. Many a company has lost money on a development contract in hopes of recovering it in production, only to have its design handed to a competitor who then takes all the gravy.)

Section VI is a statement of the binding conditions which the company is willing to accept if a contract is awarded. These are determined by management; the technical proposal writer incorporates them as given to him.

Section VII is usually a standardized statement, which often is printed in advance and inserted into all proposals at the right place.

Section VIII is prepared by the accounting department of the company, and is not the responsibility of the proposal writer.

If you, as a proposal writer, come up with the text which we have offered in this example, you will have a good skeleton for management to add to. Since this is an unsolicited proposal without a deadline, your first efforts may disappear into the company "system" for several weeks, and you may decide that you'll never hear of it again. However, it may return, with many notes and comments, and you'll then go through the whole thing again, but in much more detail. You might ask, "Why do the preliminary version at all?" The answer is that it serves as a point of departure, which enables others to add piecemeal to a basic document. Without this foundation, their ideas would have no place to come to roost, and the whole proposal would dissolve into a collection of vague unrelated suggestions.

Summary

The writing of technical proposals is a specialized branch of technical writing. It is generally more demanding and higher paid than report writing or the preparation of technical manuals. Proposals are the primary means of sales for many companies, and a good proposal writer finds himself working closely with sales, engineering, and management in his company. Proposal writing can demand long hours, and the meeting of "impossible" deadlines. It can be frustrating, and involve personality clashes between company executives. It is advantageous to learn to diplomatically sidestep such clashes. The proposal writer sees the overall operation and problems of a business from the inside, and is in a position in which one can gain experience and insight that can be invaluable in the event that he should ever decide to go into business for himself.

A proposal is a sales document, directed to a sophisticated reader. It must tell its message clearly, and briefly. It must create confidence in the company and, like all good advertisements, offer a believable promise of something the customer wants.

Proposals all follow a fairly definite logical format. Proposals written in response to invitations to bid must adhere closely to the specifications contained in the invitation. Any changes must be carefully justified under "Exceptions" in the proposal, pointing out that your company knows a better way to do it. Proposals submitted to government agencies should follow the form required in published governmental guides for proposal writing, and should in general employ the style and wording expected by government contract officers.

The proposal writer can justly consider himself to be an aristocrat in the profession of technical writing, a specialist who plays a vital part in the success of his company.

On the next few pages of this lesson, we have reproduced the pages of a small manual published by the United States Army. As you can see from the title page of the manual, it is a "Guide for Voluntary Unsolicited Proposals" -- a set of directions for the proper preparation of a proposal to be submitted to the U. S. Army Materiel Command.

Here we have a potential customer telling sellers how they should prepare proposals which are to be submitted to the potential buyer. Careful reading of this manual will probably improve your ability to write good proposals.

UNITED STATES ARMY

U.S. ARMY MATERIEL COMMAND

GUIDE FOR VOLUNTARY UNSOLICITED PROPOSALS

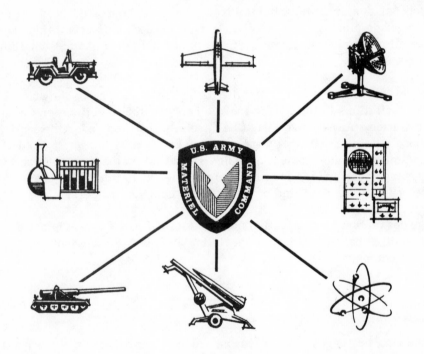

PREPARED BY
HQ, AMC (AMCRD-PS)   OCTOBER 68
PROGRAM MANAGEMENT DIVISION   WASHINGTON DC 20315

UNSOLICITED R&D PROPOSALS

Many companies have a sincere desire to apply their talents and facilities to Defense Dept. problems by undertaking research and development programs. Such programs may originate with solicited proposals or with unsolicited proposals. Let us take a closer look at this last area. Any look at the area of unsolicited proposals usually raises several basic questions such as:

(1) What are the criteria for deciding whether or not to submit a proposal?

(2) What are some of the ground rules for preparing R&D proposals?

(3) On what basis are they usually evaluated by the Army?

(4) To where should a proposal be submitted?

CRITERIA FOR DETERMINING WHETHER TO SUBMIT A PROPOSAL

In deciding whether or not to submit a proposal, industry should ask, "Are we uniquely qualified or particularly competent in this field"? Even if the answer is no, the question should be re-phrased to ask, "Are we at least as well technically qualified as anyone else?" Finally, the question should be asked, "Would it make sense for the Government to give us a contract for this work?"

If a company is certain it is uniquely qualified to tackle a problem, it is probably in a position to write a highly acceptable proposal, with a good chance for support. The following might be considered one of the rules of thumb for any proposal: If you can demonstrate uniqueness by virtue of personnel, proprietary techniques, patents, or facilities-- spell it out. It all adds up to being able to offer the Government the most economical answer to its problem. Not infrequently, companies are anxious to get into an entirely new area in order to diversify their activities. Considerable effort may then go into a proposal in vain, for unless the Company can demonstrate that its activities in another field are applicable to the problem at hand, the chances are good that

the proposal will not be accepted. Insofar as R&D is concerned, it
might be worth mentioning that the Government is not normally in the role
of assisting industrial organizations to diversify their activities.
These, then, are some of the factors that must be weighed before deciding
to submit a proposal.

GROUND RULES FOR PREPARING PROPOSALS

SUMMARY. In organizing a proposal, several considerations should
be kept in mind. First, as with any technical document exceeding 5 or
10 pages, a summary is desirable. This permits the sense of the proposal
to be gathered at once, and helps orient the reader.

INTRODUCTION.

Following the summary, an introduction may be disirable, depending upon the circumstances. Basically, an introduction is intended
to orient the reader, give him the backgound, acquaint him with the
problem, and lead him into the body of the proposal. If the summary has
already done this, it of course is not worthwhile to repeat the entire
message solely for the sake of having an introduction. In some cases,
the introduction is logical place to present additional information that
will help justify the approach you have selected. This may take the
form of information regarding some unique process or technique you have
developed which will be particularly well-suited to the problem, or a brief
explanation of the potential your proposal offers for other problems facing
the Army agency. In either case, the intent is the same--to offer evidence
supporting the proposal.

STATEMENT OF PROBLEM.

The next major portion of the proposal is usually the statement
of the problem. The intent of this section is to demonstrate your understanding of the problem. In many respects, it is one of the most difficult sections to prepare because it should present enough information
to demonstrate that you appreciate the subleties of the problem without
going into a prolonged technical analysis. Remember that the proposal

is intended to demonstrate how you would go about solving the problem,
not your ability to restate the Government's problem. In rare cases, the
statement of the problem may justifiably require supporting information,
such as a historical background or a summary of the present state of the
art. To avoid cluttering up the proposal, it may be well to extract the
pertinent facts of such sections and relegate the details to an appendix.

PROPOSED APPROACH TO THE PROBLEM.

Once the problem has been stated, the proposed approach to the
problem should be given. In many respects, this is the heart of the proposal, for it is the section that usually receives paramount attention.
A well-stated understanding of the problem, the best facilities, the most
talented personnel, and all of the other advantages that a contractor can
offer, may well be unimportant if he does not offer a logical and promising
approach.

PERSONNEL RESUMES.

The makeup and organization of the team proposed for the work
should be spelled out; and again, in order to keep the proposal unclut-
tered, resumes of the team members should be given in an appendix. Some
contractors choose the present resumes of many people in addition to those
who will be engaged on the program. To the extent that such resumes
indicate the attributes of personnel who will make supervisory or tangen-
tial contributions to the program, this may be worthwhile. On the other
hand, resumes of people who by their title or position do not appear even
remotely connected with the program could very well be construed as
padding.

WORK SCHEDULE.

In some programs, the work proposed may be scheduled in several
phases or work units. If so, it is desirable to present a section entitled
"schedule," in which the work is displayed along the projected period of
performance. For convenience and clarity's sake, a simple bar graph may
prove effective. In many instances, it may be worthwhile to include a

section on "specific qualifications". This is a useful means for presenting information on past or concurrent efforts that have specific bearing
on the proposed program. In particular, specific contracts in related
fields should be mentioned. This section may make reference to facilities or other company experience presented in an appendix.

CONTRACTUAL SUMMARY AS TO - SCOPE, COST, & TIME.

Finally, the body of the proposal should contain a firm contractual
statement summarizing the scope of work and offering to do it for a certain
sum and within a certain time. A detailed cost breakdown can be relegated
to an appendix.

APPENDICES.

Beyond the body of the proposal come the various appendices
referred to in the proposal. Even if having only general reference in
the body of the proposal, any information or photographs that will demonstrate ability to undertake the work should be presented.

BASIS OF EVALUATION BY THE ARMY.

Perhaps it would be well at this point to look to the source of
the various policies and practices that govern proposal evaluation and to
the important factors which the evaluator must keep in mind. DOD procurement practices are governed by the Armed Service Procurement Regulations,
and the basic rules for evaluating R&D proposals are spelled out in
these regulations. The item of primary importance is the technical supri-
ority of the proposal. All the resumes, annual reports, leatherbound
covers, and three-color overlays in a proposal cannot be expected to
sway the evaluator's opinion of the technical presentation. The submitter
must demonstrate (1) that he understands the problem fully and (2) that
he has a well thought-out approach which shows signs of promise if
executed as described.

On many occasions, proposals have been rejected as a result of
Submitters underrating the importance of technical superiority of a proposal.
In any proposal, revolutionary ideas that offer the possibility of sign-

ficant scientific breakthroughs are attractive. However, even old, well established approaches have won contracts. This is particularly true where the reasons for previous failure of these approaches are analyzed, adequately investigated, adn a new and promising solution proposed.

A clear understanding of the ultimate needs of the Government agency for whom the work is proposed will also assist in proposing attractive approaches. Often before preparing a proposal it may be helpful to discuss the matter with representatives of the agency concerned. If a proposal involves the development of a material that will be used in large quantities, then what will be the effect of an approach based on material in very short supply? In many instances, the availability of material may have decided effect on a given approach. Similarly, if a submitter has done his homework properly, he may find that the agency ifself has done some work on the approach the submitter has in mind, and may have some very definite ideas about it.

A technical evaluation of a proposal may sometimes be performed on the basis of the technical portion of the proposal alone. Cost informations will be deliberately denied to the technical evaluator to prevent this from influencing his decision. He is asked to rate a proposal into two broad classes, acceptable and unacceptable. He is then required to state reasons for unacceptability, and/or acceptability. Only then may cost data be brought into the picture. On the other hand, a proposal may be evaluated taking into consideration both technical and cost factors at the same time. Upon completion of the evaluation the submitter will be informed as to whether the Government is interested in supporting his proposal, or his proposal will be rejected. Sometimes the technical evaluation will be provided, other times only a statement of "no interest" or "lack of potential benefit to our research and development programs" may constitute our rejection. In any event, you may be sure that the length of the reply does not constitute the measure of the evaluation. <u>Every</u> proposal is carefully and thoroughly evaluated by highly competent personnel in the field or fields involved in the proposal.

WHERE TO SUBMIT A PROPOSAL

And now, last but not least, we come to the point that many of you have probably been waiting for since the start--"Where do I submit my proposal?"

Attached (as inclosure 1) for your information and guidance is a list of the AMC activities, together with their mission and function.

If it can readily be determined that your proposal falls into one of the commodity areas which are the responsibility of one of the listed Commands, your proposal should be submitted <u>directly</u> to the Command or agency concerned. If such a determination cannot be made, or you are in any doubt as to where to direct your proposal, send it to:

> Commanding General
> US Army Materiel Command
> ATTN: AMCRD-PS-P
> Washington, DC   20315

in triplicate.

Attached as inclosure 3 for information and guidance is a policy statement outling the conditions under which AMC will receive an unsolicited proposal for evaluation and determination of Army interest. If you are in accord with the policy outlined therein, one copy of the Memo of Understanting should be executed, signed and returned with your proposal. Local reproduction of the Memo of Understanding is authorized.

3 Incls
1. as
2. Guidelines for submission
   of proposal
3. Policy statement and Memo
   of Understanding

MISSION STATEMENTS OF AMC ACTIVITIES
HAVING A RDT&E RESPONSIBILITY

I. INDEPENDENT LABORATORIES

(Under Direct Operating Control of the Deputy for Research & Laboratories
HQ, AMC)

1. US Army Aeronautical Research Laboratories (NAS Moffett Field, CA 94035)

Accomplishes basic and applied research in the field of sub-sonic
aerodynamics applicable to aircraft, missile and other aerodynamic devices.

2. US Army Aviation Materiel Laboratories (Ft. Eustis VA, 23604)

Performs RDT&E in Army aeronautical research, exploratory and advanced
development in subsonic areas.

3. US Army Ballistic Research Laboratories (Aberdeen Proving Ground, MD
21005)

Conducts basic and applied research in (1) weapons technology and
ballistics including (but not restricted to) propulsion systems, flight
and damage mechanisms, (2) relevant areas of physics, chemistry, mathamat-
ics, and engineering (3) weapons systems evaluation, and analyses with
objective of anticipating Army's long range needs and providing necessary
technological foundation for improvement of weapons systems. Determines
vulnerability characteristics of all types of military targets, both materiel and personnel. Performs component and weapons systems effectiveness
studies and weapons concept evaluations to provide an analytical basis for
selection of weapons and design elements of components. Prepares firing
tables for all Army weapons except designated guided missiles.

4. US Army Coating and Chemical Laboratory, (Aberdeen Proving Ground, MD
21005)

Plans, manages and directs basic and applied research plus engineering
investigations in the fields of automotive chemicals, organic and semi-
organic coatings, conversion coatings, cleaners and fuels, lubricants and
related materials; coordinate the total research program of AMC in the
fields of fuels, lubricants and related materials. Participates in the
standardization and industrial preparedness program within these assigned
fields and provides technical assistance to AMC major subordinate commands
and activities on problems arising in assigned fields.

5. US Army Human Engineering Laboratories (Aberdeen Proving Ground, MD
21005)

Performs research, in and monitor the total AMC program for, life
sciences regarding human factors, i.e., capabilities and limitations of
man-machine relationships consistent with tactical requirements in the
world-wide environment and logistic considerations. Conducts associated
engineering investigations as to psychophysiological impact and specification requirements. Training personnel in assigned fields, as requested.
Assists AMC design agencies in the application of human factors engineering principles to end items and systems design.

Incl 1

6. US Army Materials & Mechanics Research Center (Watertown, MA 02172)

Manages and directs that portion of the AMC-materials research program conducted within its own laboratories, as assigned by the Director of Research and Engineering, Headquarters, AMC, including basic scientific research, and research in metals, ceramics and other materials. Coordinates the total materials research program of the AMC. Coordinates and manages a program of testing techniques, in conjunction with the quality assurance program, and executes assigned portions of standardization programs.

7. US Army Nuclear Defense Laboratory (Edgewood Arsenal, MD 21010)

Conducts research and field experiments in the nuclear weapons effects areas of initial radiation, residual radiation and fallout, shielding, and thermal radiation phenomena. Provides technical information and assistance in the fields of radiological and nuclear defense and health physics. Provides environmental monitoring and other radiological safety support. Develops radioactive waste disposal methods and shipping containers.

8. US Army Terrestrial Sciences Center, (Hanover, NH 03755)

Conducts basic and applied research in the environmental (terrestial) sciences, with emphasis on those aspects of earth physics which pertain to snow, ice, frozen ground, permafrost, and related solid earth problems in cold regions.

Performs scientific and engineering investigations pertaining to materials, facilities, systems, and military operations in cold environments.

Conducts research into methods and techniques of using various energy forms and systems to obtain information about surface and subsurface features (both man-made and natural) in all environments for engineering, military, and related scientific purposes.

Conducts such environmental and climatological research as may be required in support of the above.

Establishes general criteria, and provides advisory services for design, construction, and maintenance of military facilities in cold regions.

9. Harry Diamond Laboratories (Washington, DC 20438)

Performs basic and applied research in, (but not restricted to), the fields of radiating or influence fuzing, time fuzing (electrical, electronic, decay or fluid) and selected command fuzing; for target detection and signature analysis; and for target intercept phase of terminal guidance. Performs weapon systems synthesis and analysis to determine characteristics which will affect fuze design to achieve maximum immunity to adverse influences, including counter-counter measures, nuclear environment, battlefield conditions and high altitude and space environments. Performs basic and applied research, in support of assigned missions, or as directed by Dir/R&D, AMC, on instrumentation, measurement and simulation, on materials, components and sub-systems including electronic timers for weapons and on selected advanced energy transformation and control systems. Conducts basic research in the physical sciences, as directed by Dir/R&D, AMC. Performs basic and applied research on fluid devices and systems.

10. <u>Natick Laboratories (Natick, MA 01762)</u>

Accomplishes research and development in special aspects of the physical, engineering, environmental, and life sciences, to meet military requirements for soldier's equipment in the commodity categories of (1) clothing, foot wear, and body armor; (2) organic materials textiles, tentage and equipage; (3) subsistence and food services equipment; (4) containers and materials handling equipment; (5) POL handling and dispensing equipment; (6) field support equipment (including printing and composing equipment); (7) to develop equipment and techniques for air drop of Army personnel, supplies and equipment. As assigned, support other components of DA and the DSA, with respect to applications engineering and standardization programs, as assigned for designated commodity areas.

II. MAJOR SUBORDINATE COMMANDS

1. <u>US Army Aviation Materiel Command (St. Louis, MO 63103)</u>

Integrated commodity management of aeronautical and air delivery equipment and of test equipment that is part of, or used with, assigned materiel. Basic and applied research concerning assigned materiel development to include an item aircraft, airframe structural components, ground support equipment, wheel and brake systems, etc, gas turbine, jet engines, internal combustion radial and horizonally opposed aircraft engines, and aircraft hydraulic pumps, starters, etc.

2. <u>US Army Electronics Command (Ft. Monmouth, NJ 07703)</u>

Responsible for the research, development, procurement, maintenance support and supply of communication and electronic equipment and systems. Principal interests are communications, communications security, electronic warfare, aviation electronics (avionics), night vision, combat surveillance target acquisition, electronic intelligence, photographic and microfilming, air defense electronics, identification--friend or foe (radar) systems, automatic data processing, radar, meteorological and electronic radiological detection materiel (except fire control, radar, computers, and closed circuit computer systems integral to a weapon system), electric power generation equipment, and related assigned special purpose and multi-system test equipment, component parts and materials.

Conducts continuing research (relating to the missile program) in the fields of missile electronics warfare, missile vunerability, and missile surveillance. Coordinates the missile electronics counter measures efforts of the US Army.

3. <u>US Army Missile Command, (Redstone Arsenal, Huntsville, AL 35809)</u>

Exercises integrated commodity management of assigned materiel, e.g. free rockets, guided missiles, ballistic missiles, target missiles, air defense missile fire coordination equipment, special purpose and multi-system test equipment, missile launching and ground support equipment and missile fire control equipment. Conduct a basic and applied research program with respect to assigned materiel and such other research projects as may be assigned by HQ, US Army Materiel Command.

4. **US Army Mobility Equipment Command, (St. Louis, MO 63166)**

R&D Effort is accomplished through (MERDC), US Army Mobility Equipment Research and Development Center, (Ft. Belvoir, VA 22060)

Accomplish basic and applied research and development with respect to assigned items of equipment, e.g. rails, marine and amphibious equipment, construction, electric power generating, bridging and assault stream crossing, fire fighting, prefabricated buildings, waste disposal, heating and air conditioning, night vision, camouflage and concealment, mine warfare, barrier and intrusion detection, demolitions, water purification, petroleum storage and distrubution, industrial engines, and land navigation.

5. **US Army Munitions Command, (Dover, NJ 07801)**

    a. **Ft. Detrick, MD 21701**

Conducts research, exploratory development, engineering development, testing and evaluation, and associated activities in the DOD offensive and defensive research and development biological programs.

    b. **Edgewood Arsenal, MD 21010**

Conducts research and development in the offensive and defensive aspects of physiologically active chemical compounds, chemical degradation of materials, aerosol physics, basic research in the mechanism of action of body systems; and carry out research and development on smoke, incendiaries, and non-electronic antisurveillance techniques.

    c. **Frankford Arsenal (Philadelphia, PA 19137)**

Operates a commodity center for small caliber munitions, cartridge activated and propellant actuated devices. Conducts research with respect to assigned commodities including specialization in optical materials and technology, metallurgy, of non-ferrous and reactive metals, materials degradation, power transmission fluids for small control mechanisms and special synthetic lubricants, materials and technology pertinent to miniaturization of ammunition. Performs design and development for assigned commodities and for artillery shell metal parts and cartridge cases.

    d. **Picatinny Arsenal (Dover, NJ 07801)**

Operates a commodity center for nuclear munitions; radiological materiel; and artillery and mortar ammunition, and non-chemical and non-biological bombs, mines, grenades, demolition devices, explosives and explosive devices, propellants, pyrotechnics, boosters, JATO's, and rocket and missile warhead sections. Conducts research with respect to assigned commodities including specialization in the following fields; plastics and adhesives, solid and liquid propellants, dynamics of materials, non-metallic materials other than rubber, greases, lubricants, corrosion preventives, and fuels. Performs, design and development for assigned commodities and for small caliber munitions fuzes in support of Frankford Arsenal; for impact fuzes, inertial fuzes, safing and arming devices and barometric devices for assigned commodities and in support of MUCOM commodity centers. Operates the DOD Plastics Technical Evaluation Center.

6. **US Army Tank-Automotive Command, (Warren, MI 48090)**

Conducts applied research, development and engineering of tank-automotive vehicles and components for the Army and other Defense Department activities. Typical R&D Programs include power train components, suspensions, trucks, trailers, tires, new concepts for mobility, armor concepts, weapons systems for tanks and combat vehicles, and all types of vehicle engines and accessories.

7. US Army Test and Evaluation Command (Aberdeen Proving Ground, MD 21005)

Plan and conduct engineering and service tests and evaluations; support engineering design, production and post-production tests, and participate in troop test planning. A number of installations and activities are involved, such as: Dugway Proving Ground, White Sands Missile Range, Aberdeen Proving Ground (D&PS), Army Electronic Proving Ground and Yuma Proving Ground.

8. US Army Weapons Command, (Rock Island Arsenal, IL 61201)

Conducts research, design and development of artillery, artillery mounts, recoil mechanisms, carriages, loaders, hand carts, arms racks, target materiel (except aerial drones), common tools, tool sets, and shop equipment.

Performs research, design and development of individual weapons, machine guns, grenade launchers, secondary armament for combat and tactical vehicles, aircraft armament subsystems (gun type), spotting weapons, mounts and pods for mission weapons, links, linkers, delinkers. Is also responsible for research, design and development of mortars, recoilless rifles, cannon assemblies and components.

9. US Army Sentinal Logistics Command, (Redstone Arsenal, AL 35809)

Provides mission essential logistic support to the Sentinal System except for nuclear munitions and auxiliary equipment.

GUIDELINES FOR SUBMISSION
OF A PROPOSAL

There is no specifically prescribed format for a submission of an unsolicited proposal for a research or development contract. However, the following information is offered to assist you in preparation of your proposal.

A submittion generally consists of:

a. A cover letter stating clearly and briefly the objectives and scope of your proposal, any past experience or studies that are pertinent to the subject at hand, and a statement of total cost.

b. Inclosures

(1) The proposal itself. This should contain a clear and full discussion of proposed work. It is preferable that it be securely stapled together or inclosed in a suitable folder or cover. A suggested outline is attached. Naturally each proposal may be varied to suit the individual requirement of the submitter.

(2) Detailed Cost Estimate

(a)	Direct Engineering (hours)	$ XXX
(b)	Direct Engineering (dollars)	$ XXXXXXXX
(c)	Engineering Overhead	$ XXXXXXXX
(d)	Direct Charges	$ XXX
(e)	Total Manufacturing Cost	$ XXXXXXXX
(f)	General & Administrative Overhead	XXX
(g)	Total Cost	$ XXXXXXXX
(h)	Fixed Fee (if on a total cost plus fixed fee basis)	$ XXX
(i)	Total cost plus fixed fee (if applicable)	$ XXXXXXXX

(3) Special or Proposed Contract Terms and Conditions (if applicable).

(4) Any other data considered pertinent.

Proposal should be submitted in triplicate to appropriate major Command or agency concerned with the work proposed, or, if this cannot be readily determined, to:

Commanding General
US Army Materiel Command
ATTN: AMCRD-PS-P
Washington, DC 20315

DEPARTMENT OF THE ARMY
HEADQUARTERS UNITED STATES ARMY MATERIEL COMMAND
WASHINGTON, D. C. 20315

STATEMENT OF POLICY

Conditions Under Which The Army Will
Receive and Evaluate an Unsolicited Proposal

The Army has a continuing interest in receiving and evaluating proposals containing new ideas, suggestions, and inventive concepts for
weapons, supplies and equipment. However, Government personnel and contractors are constantly engaged in research and development activities,
and the substance of your proposal may already be known to Government
employees or contractors, or may even be in the public domain. For such
reasons we have found it to be desirable, when receiving proposals for
evaluation, to insure that the persons submitting them are aware of the
conditions under which they will be considered by the Army.

It should be understood that the receipt and evaluation of a proposal by the Army does not imply a promise to pay, a recognition of novelty or originality or any relationship which might require the Government to pay for use of information to which it is otherwise lawfully
entitled. However, the Army has no intention of using any proposal in
which you have property rights without proper compensation.

If you agree to the above terms, the Headquarters United States Army
Materiel Command will be glad to have your proposal referred to our cog-
nizant personnel for consideration and evaluation and will report to you
the extent of our interest. Any material which you submit, will in the
evaluation process, be restricted to only those persons having an official need-to-know for the information for purposes of evaluation.

Please sign one copy of the Memorandum of Understanding (reproduced
on the back of this Policy Statement) if you assent to the conditions
stated, and return it to the address indicated by our letter for file.
You may keep one copy of the Memorandum for your own records.

Incl 3

AMCRD-SS-P FL 8,
20 Jun 67

MEMORANDUM OF UNDERSTANDING

Place_____

Date_____

        The undersigned acknolwledges that this date he has, on behalf of
(himself, or _____) made a disclosure of an in-
                (Company or Corporation)
ventive proposal to the Department of the Army relating to

_____

_____

It is understood that the Department of the Army has accepted the above
proposal for the purpose of evaluating it and advising of any possible
Army interest, provided that the acceptance to determine such interest
does not imply or create a promise to pay; an obligation to give up any
legal right or to assume any duty; a recognition of novelty, originality
or priority; or any express or implied contractual obligation or other
relationship such as would render the Government liable to pay for or to
give up any legal right or assume any obligation for any disclosure,
evaluation or use of any information in the proposal to which the Government would otherwise lawfully be entitled.

                        _____
                                    (Signature)

                        _____
                                (Printed or Typed Name)

                        _____
                            (Title or Position, if appropriate)

Incl 3

AMCRD-SS-P FL 8
20 Jun 67

TEST TC-7

TRUE-FALSE QUESTIONS

1. Companies which sell through retailers to the public use the most technical proposals .... _____

2. Proposals are much used in setting up research and development contracts with governmental agencies .... _____

3. A proposal is in effect a part of the selling effort of a company .... _____

4. A proposal is a legally binding document .... _____

5. The cost of preparing a proposal is a part of a company's overhead .... _____

6. The expression "or equivalent to" in an invitation to bid means that all bidders have an equal chance to get a contract .... _____

7. The "Exceptions" section of a proposal details any desired deviations from specifications spelled out in an invitation to bid .... _____

8. The proposal writer begins his work by first writing the letter of transmittal. . _____

9. The proposal writer should write the "Summary" last .... _____

10. The "Statement of the Problem" in a proposal, in response to an invitation to bid, need not be much concerned with the specs outlined in that invitation .... _____

MULTIPLE-CHOICE QUESTIONS

11. The "Table of Contents" in a proposal is usually followed by a
    1. List of illustrations   3. Summary   5. Statement of the Problem
    2. Letter of transmittal   4. Introduction   _____

12. The alphabetical index is placed
    1. Right after the letter of transmittal   4. Right before the summary
    2. Just before the list of illustrations   5. Just before the appendix
    3. At the very end of the proposal   _____

13. The largest proposal-receiving customer is
    1. General Motors   4. The U. S. Government
    2. Standard Oil   5. The State of New York
    3. Bank of America   _____

14. The key section of a proposal discusses
    1. Qualifications of the company   4. The Summary
    2. The basic plan for the solution of the problem   5. The Exceptions
    3. Production planning   _____

15. An invitation to bid
    1. Means that only one company is asked to submit a proposal
    2. Means that one company has the edge over others in getting a contract
    3. Is usually sent to those companies which are believed qualified to do the job
    4. Is sent only to companies which have done business with the customer
    5. Is a guarantee that the company receiving it will get the contract   _____

16. The style of a proposal is usually
    1. Quite formal
    2. Highly informal
    3. Similar to that of an interdepartmental memo
    4. In the first person
    5. Unimportant

17. The proposal writer usually spends the most time writing
    1. The letter of transmittal
    2. The summary
    3. The technical theory discussion
    4. Exceptions
    5. Qualifications of the company

18. A technical writer is seldom concerned with
    1. The frontispiece
    2. The summary
    3. The introduction
    4. The technical theory discussion
    5. The letter of transmittal

19. The "Summary" of a proposal is usually
    1. The first thing written
    2. The first thing read
    3. Not read by a casual reader
    4. Not prepared by the technical writer
    5. Given to the writer by his superior at the start of his writing job, for his guidance in preparing the proposal

20. In general, a proposal may be described as
    1. Having the same relation to a company as a personnel record has to an employee
    2. A document prepared primarily for the company sales department
    3. Primarily a guide for the engineering and production departments
    4. The major means for getting contracts for many companies
    5. Of primary importance to companies who sell through jobbers

WRITING EXERCISES

21 thru 23: Practice in proposal writing is best obtained by writing proposals. It is difficult to set up academic exercises because of the volume of information that is needed. The following ten situations (A thru J) are offered as typical of products and ideas in which you can fill in the details. Select two of these ten situations, and write these 2 proposals as a part of this test.   In writing up these proposals, follow the outline given in the lesson. Each section may be quite short, depending on what information you feel is needed. The final result of your work will probably be like our example -- that is, will be the foundation for a complete proposal to be developed in conferences with other people in your "company". Remember, select only 2 from the following 10. Identify those selected by the identifying letter used below.

You will note that in each of these items you must use your own creative ability and technical knowledge to decide how the job will be accomplished. You must be the inventor as well as the proposal writer here. Thus this lesson can be considered as practice not only in proposal writing but also in the most important of all the functions of the engineer -- invention.

A. Write a proposal to a toy distributer for a talking doll which contains a radio receiver tuned to a short-range transmitter, so that a mother or someone else can make the doll apparently carry on a conversation with a child. In this proposal, discuss the psychological aspects of such a toy, as well as matters of construction and cost.

B. Write a proposal to the U. S. Department of Defense for a talking light beam communication device consisting of a binocular frame, one side of which contains a light modulator/ receiver unit, and the other side of which serves as a sight.

   In Proposals C thru J, decide on what kind of customer the proposal should be addressed to. You may want to make some of these inter-departmental proposals, written by a department such as sales or engineering to the management of your own company. While such internal documents are really technically memos, they can have all the attributes of a formal proposal intended for submission to an outside organization.

C. Write a proposal for a rifle scope with illuminated crosshairs.

D. Write a proposal for an electronic organ which plays from player piano rolls.

E. Write a proposal for a pair of tongs to pick trash in the yard (grass cuttings, weeds, papers, etc.)

F. Write a proposal for the most compact possible 45 rpm record player.

G. Write a proposal for a flashlight which uses a small hand-operated generator instead of a battery.

H. Write a proposal for a 9 volt generator for transistor radios, which is turned by a wound-up spring motor. Assume one winding to be good for 15 minutes playing.

I. Write a proposal for a market cart which is supported by an air-cushion instead of wheels.

J. Write a proposal for an electric typewriter with special keys that make the machine type out whole words, or phrases like "Very truly yours,".

LESSON TC-8
Writing Instructional Materials

Introduction

All technical writing has instruction as its ultimate goal. Operation and service manuals teach the users of equipment how to handle it. Proposals give information which enables contracting officers to make awards. Reports give the necessary information to company executives so that they can make decisions. Even articles written for popular magazines enable people to learn, for the benefit of job or hobby, or just for the pleasure of knowing about science and technology.

There is another kind of technical writing, however, which is expressly aimed at education in a broader sense than the highly specialized technical manual. In this kind of writing, the writer produces training manuals, syllabuses, and textbooks. In this activity the writer becomes a teacher, and so his veiwpoint and orientation must be that of a teacher. He is not writing for executives who want information about a specific subject, or for trained technicians who want to know how a particular device is operated. His job is to furnish the mind of his reader with a body of knowledge, and his approach must be to build up that knowledge in a way and at a rate which the average human mind can take.

Before one can write educational material, he must have an understanding of the process of education itself. To gain such an understanding is a big order. Colleges devote years of study to education, and grant teaching credentials and academic degrees up to the Ph D which are supposed to indicate skill in teaching. There are many theories as to how a person learns, and there have been and are bitter controversies with regard to educational philosophy and teaching methods. The writer of this lesson, has been through the mill in education, having taught in junior high schools, high schools, junior colleges, state colleges and universities, private liberal arts colleges, and the California Institute of Technology. This exposure has helped him formulate certain principles of teaching, both in the classroom and in the preparation of written materials. These principles, plus well-known methods expounded in texts on education and taught in education classes, form the basis for the discussion which follows.

How Do We Learn?

If we had the complete answer to the foregoing simple question, education would be greatly simplified. Teachers, psychologists, and educational experts spend much of their time debating this question, and to date there is no complete agreement on an answer. We can, however, make some useful observations about the general process of learning, from personal experience and observation. We are born with a certain basic "knowledge." Like the animals, genetics provides the neural connections in the brain to enable us to carry on immediate vital functions. We know how to breathe, swallow, and excrete, and we learn how to control our bodies in walking, using our hands, and seeing, without being taught by others. We also share with animals the ability to learn by observing and imitating. Perhaps the most important skill acquired in this way is speech. Most other skills are also best learned by observation and trial. In technical education, this fact is applied by the demonstration lecture followed by laboratory work. "Learn by doing" is accepted as the best way to develop a skill.

In seeking an answer to the question, "How do we learn?", it is helpful to consider just what we learn, and how this may be categorized. The content of a person's mind represents pretty much what he has learned in his lifetime. This can be classified into three basic categories:

1. Factual knowledge

2. Technical skills

3. Attitudes and policies

By factual knowledge we mean information -- Ohm's law, the value of π, the number of foot-pounds in a BTU, the population of New York City, your social security number, the Lord's prayer, the rules for playing Bridge, the names and dates of history, the words of a popular song.

We draw upon our store of factual knowledge everyday, for reference, in coming to a decision. Decisions are based on comparisons. For example, is the price of something higher or lower than that of another equivalent item, or of the same thing at another store.

Factual knowledge is memorized. That is, it is repeated to us a sufficient number of times to create the peculiar neural patterns that constitute memory. We learn a phone number by repeating it several times, or better, by using it so often that presently we no longer have to look it up.

Memorizing uses the capacity of the brain, and it is possible to cram so much information into the memory that the addition of a new fact will literally cause you to lose conscious track of some other bit of information. For this reason, we should ration the number of things we remember, and depend on auxiliary "memories" such as notebooks for facts we use infrequently. Which things should we remember, and what should be relegated to notebooks? This question is answered rather automatically. If we need to use a given fact often enough, the mere act of looking it up will presently put it in the active memory. The decision of whether to memorize or not is thus often made automatically without volition.

Technical skills are a special kind of knowledge, needed in order to do something. Some skills, like the handling of tools, are acquired and improved in the act of application. Other skills, like the integration of mathematical functions, can be developed by thinking them through and contemplating the nature of the act, as well as in their actual performance. Skills range all the way from the ability to hit a nail squarely with a hammer to the ability to make logical deductions. Many skills involve bodily dexterity, like dancing and expert soldering. Others are purely mental in their application.

Attitudes and policies provide the motivation for the use of information and the exercise of skills. Attitudes are formed by observation of other peoples attitudes and the consequences of these attitudes, good or evil. When you believe that "honesty is the best policy" you have formed an attitude which will determine the use to which your knowledge and skills are put.

Attitudes can be formed by listening to a sermon, but more often they are the result of observing another person as a good or bad example.

How are the foregoing three categories taught?

Factual knowledge is imparted by repetitive exposure, i. e. , by drilling. And technical skills are developed by practicing over and over, and such learning can be taught by ordinary drill or by the use of a teaching machine. If you are teaching a subject like a foreign language, which is pure information, you will have the students repeat words and sentences until the language becomes a basic part of the students' automatic thinking. If you are teaching how to play the piano, you will require the student to repeat finger exercises until the correct movements become habitual. If you are teaching soldering, you demonstrate the action, explain it, and have the students perform the functions repeatedly.

The third category -- the development of attitudes -- is far more subtle. Attitudes cannot be created by a teaching machine; this process requires a human teacher in full-time attendance. This teacher cannot drill the class, or try to make them learn by doing; the consequences of practicing a bad policy might be too serious! The teacher can do his job best in the area of attitudes by serving as an example of the application of the attitudes he wishes to expound. The first teaching was visual and/or verbal and person-to-person. As the American educator Horace Mann once put it, the basic requirements for a school are a teacher, a student, and a log for them to sit on. As the subjects taught became more complex, it became necessary to record the instructions in writing, both for the teacher's reference and for the student to study between class sessions. These writings were the first "instructional materials," and from them as ancestors have come the whole spectrum of syllabuses, textbooks, compendiums, and other forms of writing designed to impart new knowledge.

Let us consider some of the kinds of writing done for instruction, as to purpose and style.

The Syllabus

Any course of instruction must start with an outline, which lists the topics to be covered. When an outline is expanded by stating briefly the facts to be discussed under each topic, it becomes a syllabus.

A syllabus is not normally used by a student to learn a subject, because it lacks detailed explanations and examples. It is like a skeleton, without any meat on the bones. When a student takes lecture notes, he essentially is writing his own syllabus of the course. The information he records has meaning to him because it serves as a cue to his memory, in recalling the details supplied by the teacher in lecture and discussion. Teachers often prepare syllabuses (the Latin plural, syllabi, may sound better to you) to save students time in note-taking, and to allow them to listen to the content of the lectures rather than becoming frantic note-writers.

College students often find it profitable to take careful notes in a lecture class when no notebook or syllabus has been prepared by the instructor. These notes are then often published in ditto or mimeograph form and sold to the other members of the class, in time for use before final examinations. The profit from this type of literary enterprise arises from the fact that the student publisher has an excellent salesman in the person of the instructor, whose hard sell is the final exam. In a large lecture class the writer-publisher may reap a tidy profit in the hundreds of dollars, and have as a side-benefit an A grade in the course, because in writing the syllabus he must of necessity learn the subject very well. The writer of this lesson engaged in such an operation years ago at the University of California, except that he prepared a 176-page textbook, derived from the professors notes and many other textbooks. This book became the standard text for the course for seventeen years, and brought in a small but helpful income long after he departed from Berkeley.

A very successful publishing venture in syllabus writing is found in the several series of "college outlines" published for the use of students in colleges and universities all over the country. The Scaum Publishing Co. of New York, for example, puts out many volumes in its series, covering science, mathematics, and many liberal arts subjects. These books contain essential theory, brief explanations, and derive their greatest benefit from hundreds of worked-out examples illustrating the theory covered. The examples are intended for the use of students who learn theory by applying it, but many professors have discovered in them a gold mine for examination questions, saving them the time and trouble of making up their own and working out correct solutions. An interesting opportunity is offered to students whose professor engages in the foregoing practice, for if they study the examples thoroughly, they just may hit the jackpot in the final exam!

The preparation of syllabi (we are now switching to the Latin plural) and outlines such as those mentioned offers little challenge of a literary nature. Discussion is kept to a bare minimum, and the main concern is the selection of topics, the correctness of the theory presented, and the offering of problems that are good illustrations of theory application.

The publishers of outline syllabi, like the Schaum company, usually pick their authors from major colleges, since this is the best way for them to insure a good selection of subjects (and names which help sales). If there is a specialized course taught locally, however, there is nothing to prevent you from working out an arrangement with the instructor to divide the profits, and then preparing and printing your own College Outline. You can even do this without the cooperation of the instructor, by just enrolling in the course and writing up your own notes.

If you should find yourself teaching a night course, you have the best of all opportunities to prepare a syllabus for your captive audience. When you do this, you will also find that the careful thought and planning required to write the syllabus will have a very beneficial effect on the quality of your teaching. In addition, you will find that the teacher who also writes his own text material commands the respect of students, fellow teachers, and the school administration.

The Compendium

This formidible sounding word simply means a publication in which one has gathered together all the information he could find on a given subject. A compendium is not arranged in an order logical for teaching, but is arranged according to subject matter. It is a reference work, to which both students and practitioners turn for information. The biggest compendiums are the encyclopedias, which profess to list all human knowledge in 57 volumes (or whatever the number is).

The writer of a compendium (or of a part of such a work) should first be an authority on the subject. Next he should be able to express himself in an efficient condensed style, since space is usually at a premium in a book crammed with information. The style of writing is usually rather formal, but clarity should never be sacrificed for supposed elegance. The typical reader of a compendium is a person acquainted with the subject, who wants to learn some special facts. In writing a compendium you can assume considerable background knowledge, or at least above-average intelligence.

Your best examples of compendium writing are to be found in the major encyclopedias. Articles written for some types of journals, such as the technicians journals mentioned in another lesson in this series, often approach good compendium style.

Writing for a compendium is usually done by invitation of the publisher, or by an editor-in-chief who coordinates all of the articles prepared by different writers. If you are just starting as a writer, or as a technical practitioner in your field, you may have quite a wait before an invitation will arrive in the mail to prepare the article on transistors for the Brittanica. But this form of instructional material must be written in many fields, and the chance to assist in a compilation of, for example, company procedures, may be not at all remote.

The Textbook

The king of all educational writing is the textbook. A good textbook serves as the basic source of subject material and examples for a course. With such a book, an instructor can prepare a course plan by doing little more than dividing the number of pages by the number of weeks in the school term. If he knows the subject well, the teacher can even get through the course without cracking the book himself -- although this practice may produce some interesting discussions when he disagrees with the text.

Textbooks are seldom written cold by writers, no matter how authoritative their knowledge or great their writing skill. A textbook reflects the educational approach and method of a course, and it can seldom be written except by an instructor who has taught the course for some time.

The usual sequence of events in the preparation of a textbook is the following:

1. Before the course is given for the first time, the teacher prepares an outline of subjects to be taught. He may at this time select a ready-made text, in which case he reaches a dead end as far as the preparation of a new book is concerned.

2. The consciencious teacher will prepare and print for the students a course syllabus in advance of the first offering of the course. A lazy teacher will not bother to do this, but will force the unfortunate students to take notes from the blackboard at the same time that they are trying to follow the details of the lecture.

3. When the course is given for the second time, the syllabus can be rewritten in expanded form. At this time, topics arranged under outline headings may be changed to separate chapters with names. The material is still a syllabus, but the change in format also makes it a kind of embryo textbook.

4. During the second, third, and fourth times the course is taught, the instructor should start budgeting some of this time toward writing up each chapter in more expanded form, as a rough draft of a future textbook. Each chapter at this time might consist of twenty or thirty pages of mimeograph material, and these chapters would be issued separately during the course.

5. After the rough-draft text has been issued and used for a year or so, the instructor will face the decision of formalizing it into a real text, with illustrative examples, problems at the end of each chapter, index, and all of the other adjuncts required of an academic publication. The wise professor will do this work but resist the enticements of book publishing firms, keeping it in mimeo form for one more year. In this way he will be able to "de-bug" his work -- improving explanations, correcting errors of fact, preparing a good answer-book for problems, and otherwise fixing things up to save himself embarassment later on when everything is frozen in typeset pages.

6. If all of the preceding steps have been taken, the day will finally dawn when the writer finds himself signing a contract with a major textbook publisher. Then the fun begins. First he must prepare and send off a bulky manuscript. Next he must gather all of the illustrative materials -- the sketches, diagrams, graphs, and charts -- for the publishers artists to redo in professional quality.

Then, after a wait of a few weeks or months, he receives long strips of paper with several pages printed on each -- the galley proofs. His job is to read these and note all typographical errors. The process of proof reading contains two main sources of trouble. First, in reading your own material you are apt to miss typographical errors, since one tends to see what is expected. Next, and more dreadful, you will discover many places where you would like to make a change in your own writing. This temptation can be very costly, and some publishers charge the authors for these kinds of changes, deducting the cost from future royalty payments. The only defense against this trouble is to do your reviewing and editing very carefully in the manuscript before it is set in type. Here you reap the reward of preliminary editions in mimeograph form.

In proof-reading you should get other people to read the galleys also. Wives, sweethearts, and other relatives are prime victims, for they can often spot mistakes even when they can't understand what the book is about at all. Professors can often sucker in their students for proof-reading, by making them think it is an honor to be so chosen. (The professor can also salve his own conscience, if any, by rationalizing that the student also profits by learning in the process. The student doesn't learn much, of course, because he is thinking about errors rather than the subject matter.)

In order to sell textbooks, the writer must appeal to several judges. In the last-but-not-least category we have the students who will endeavor to learn by reading it. They have little voice in the initial selection of a book, but in the long run, the continued use of a text depends on how effective it is with the students. Clarity, accuracy, and the use of a style and language appealing to the age group it is written for, determine the success of a text. From the standpoint of a teacher, the best textbook is the one which makes his work easy. If he has to interpret the book in detail, he will rather quickly discard it for another which appeals more to the students.

The actual selection of texts is usually the responsibility of the teacher (in a college) and the school board (in high schools and elementary schools). In the latter case, authority is usually delegated to a committee of teachers, who generally ask the opinions of working teachers who will use the book in class. Texts in the social sciences must often meet requirements that are political in nature, as well as the basic tests of accuracy and good writing. The struggle between the political right and left is never stronger than in the field of teaching, because both sides realize that control of the ideas presented will determine to a great extent the attitudes of the next generation. Politics is not a big issue with technical books, although it is possible even here to give a subtle bias (e.g., that the Russian, Popoff, instead of Marconi, really invented radio). The writer of a technical text should take a strictly neutral stand on non-technical matters, for in this way he is most likely to get both factions in the political controversy to turn to him as a compromise when they have stymied each other on politically-oriented selections.

When a teacher sits in judgment on a selection of textbooks, other factors peculiar to his work come into play. To the teacher, a text is several things.

First of all, of course, it takes over part of the job of teaching. In the ideal textbook, the teacher can assign homework reading and problem solving, and let the book do all the lecture work, confining his time to listening to recitations and answering specific questions. We have already noted that a well-organized textbook can serve as the outline of a course, saving the teacher the time and effort of planning. The examples given at the end of each chapter are also important to the teacher, because they save him the trouble of making up his own. A good teachers' answer book, with detailed solutions to each problem, is an irresistible lure, as it saves a great deal of teacher "homework" and may spare the instructor the embarrassment of getting a wrong answer, or finding out that he doesn't know how to work out some problem.

A good textbook can teach the teacher as well as the student. A harried teacher assigned to a subject not in his field can use the book to keep one jump ahead of the class, and turn out a creditable job of instruction. Indeed, it often turns out that one can teach better something just learned and still fresh to him, than topics which have been worked over so many years that they are old and dull.

How does one organize and write a textbook? There are almost as many ways as there are books. A text on an advanced technical subject will organize its chapters according to subject matter, with each one serving as a quasi-independent treatment. Other textbooks will build up the knowledge of the reader chapter by chapter, requiring that all chapters be used. Texts in which one or more chapters can be skipped without impairing the understanding of other parts are appreciated by teachers because they allow more flexibility in using the text for courses of varying lengths. You can make your text more valuable (and more likely to be adopted) if you can say something like the following in your introduction:

"For a six-week survey of the subject, use Chapters 1, 2, 3, 6, 9, and 13. For a ten-week quarter course, meeting three times a week, add Chapters 4, 11, and 15. For a fifteen week semester, also add Chapters 5, 10, and 16. For a full year course meeting three times a week, use all chapters."

The number of chapters in a textbook may be quite important to the user. Traditionally, a textbook is divided into eight to twelve chapters averaging 30 pages in length. Ideally, there should be one major subdivision covered in each chapter. In his work of organizing a text, Kinematics, from the manuscript prepared by the late Professor Howell Tyson of California Institute of Technology, the writer of this lesson used only six long chapters, each treating a single topic in considerable depth. At the other extreme, he divided his elementary Introduction to Engineering into 32 short "units" of 6 to 10 pages each because of the large number of subjects treated in a survey style.

If a textbook is prepared for a course taking, say, 15 weeks, then it may be written so that there are 15 chapters each constituting one week's instructional material. If one chapter represents several week's work, then the writer might consider dividing it into sections constituting sub-chapters, each of which corresponds to a week of class instruction.

Chapters are usually subdivided into numbered sections, which may be listed in the table of contents. Subdivision makes it easier for a teacher to make reading assignments than when page numbers must be given, and simplifies cross referencing in the book. When a text is used as a reference, or source of technical information, a listing of sub-sections in the table of contents facilities the job of looking up a given topic.

The main parts of a textbook are the following:

1. Preface. This section serves a number of purposes. It may state the author's reason for writing the book, or the specific needs which he feels it fills that other texts do not. The preface is written for the teacher rather than the student and is in some ways a professional confession addressed to the writer's colleagues. Credits and acknowledgements of help received from others are given in the preface. For example: "The Author desires to express his deep appreciation for the assistance given by Professor Ossip Splopsizzle, Dr. Chumley P. Bilgewater, and Umbilico Q. Influenza, for their assistance in compiling material used in this volume, and to his many other colleagues without whose help the book would not have been possible. He would also like to acknowledge the patience of Miss Ophelia Eyesore who typed the manuscript, and the unvarying devotion of his wife who spent many sleepless hours reading the proofsheets. In addition. . ."

In the preface the writer may also express his philosophy in writing the book, and give suggestions to teachers as to how it should be used. Also, the preface is the proper place for recommendations concerning the use or skipping of chapters for short courses, and tips as to which sections may require special explanation in the classroom. Here the writer should spell out what other courses the students must take as prerequisites for a class using the book, and the nature of additional study which the text prepares for.

2. Introduction. The introduction is really a preliminary chapter of a book. Its purpose is to survey the whole subject, outlining the contents of the book and relating this to other related subjects. For example, a text on kinematics might have the following in its introduction:

"The science of mechanics is composed of three subdivisions: statics, which studies forces on structures which do not move, kinematics which studies the motion of machine parts as abstract geometric elements, and dynamics which studies the motion of machine parts while taking into account forces and the energy changes involved. Statics is the result of combining force with geometry, kinematics is the wedding of time with geometry, and dynamics brings together all three elements: force, time, and geometry."

The introduction is addressed to the student as well as the teacher. It is unfortunate that many student readers skip the introduction and just start in reading the first chapter, which usually starts the detailed study of the first subject considered. In order to prevent this tendency which deprives the student of a necessary preliminary overview of the subject, the writer of this lesson recommends that a formal introduction be omitted, but that Chapter I contain the same information.

3. The main body of the text consists of a group of chapters which discuss the various devisions of the subject treated. The text on kinematics previously mentioned has the following chapters:

1. Machines and Their Mechanisms. This introductory chapter describes linkages, cams, gears, and other mechanisms and the mechanical effects which each produces.

2. Vectors and Machine Motion. The analytic part of the book starts out with a general discussion of graphical vectors, and their use in the analysis of velocity and acceleration of machine parts.

3. Velocity Transfer. The use of vectors in the study of velocities in linkages is studied in detail.

4. Acceleration in Machines. In this chapter the discussion started in Chapter 2 is continued in detail.

5. The General Equation of Motion. This chapter combines the discussions of Chapters 3 and 4 and presents a very complete theory of the motion of machine parts.

6. The Gyroscope. Out of the many applications of the general theory developed in Chapter 5, the gyroscope is selected as the one which serves as a good example of how theory is reduced to the level of practice, and also is of very great importance in the guidance and control systems of missiles, satellites, and spacecraft, as well as less sophisticated things such as airplanes and ships.

In dividing the subject matter of a text into chapters, the author must consider the several factors mentioned, such as number of subjects, the course in which the book will be used, and the level of student preparation.

4. Auxiliary sections of the book.

Many textbooks contain appendices. These are essentially expanded footnotes, i. e. , material of a detailed nature which is presented away from the main discussion in order to avoid interruption of the primary train of thought. Typical appendix material would be the derivation of an equation used in the text, or the description of experiments leading to a conclusion used in the book.

Kinematics, the book discussed earlier, has four appendices: a list of and the meaning of the notation and symbols used in equations, a discussion of how graphical scales used in accleration vectors are worked out, the derivation of the normal acceleration equation in three dimensions, and a detailed derivation relating to relative angular acceleration in three dimensions.

Almost all textbooks contain an index. An index is an alphabetical list of topics, and names of objects or people, given with the page numbers where they appear. An index should have at least a hundred or more entries to cover a major technical book. The major discussion for a given subject should be indicated in the index with the page number printed in boldface type. Discussions covering several pages would be indicated by giving the starting and finishing pages, as 273-294. A long discussion starting on a given page might be indicated by the number and a dash, as 75--.

The index of a book is prepared by the author in the following way.

He reads through the book, page by page, with a stack of 4x5 slips of paper or file cards at hand. When he finds a topic that should be in the index, he writes it on the top of a slip and puts the page number in. As the slips are accumulated, they are put in a file box in alphabetical order. (This is a good job for the author's wife.) When he has gone through the whole book, all slips having the same topic on them are combined into a single slip. The whole alphabetical pile of slips is then typed out as an index. It is usually wise to preserve the original box of slips for future changes. In any major change, however, the page numbers will all be different, so that the job will have to be done all over again.

A special kind of index is known as a concordance. This is an alphabetical list of every noun and verb used in a book, with the exception of such verbs as to be or to have. In the past, Biblical scholars have spent years of time preparing concordances for particular editions or printings of the Bible. More recently, IBM cards were used to make the most complete and accurate concordance ever prepared. Every noun and verb was placed on an individual card, with a page number. The several million cards were then run through a card sorter, set to select all cards having first letter, A. The remaining cards were then run through the sorter, set for B, and so on through the alphabet. Each of the 26 piles was then rerun, with the sorter set for A, B, C, etc., for the second letter. By a continual repetition of this process, the millions of cards were finally arranged in alphabetical order. The cards were next fed through a line-printer, which produced the page copy for the concordance, entirely without human work, save for the initial punching of the cards.

Much can be written on the subject of textbooks. Obviously the style, vocabulary, and general treatment of the subject depends on the age, background, and general preparation of the students for whom it is intended. Study of the thousands of textbooks which are used in schools and colleges will give examples, both good and bad, of textbook writing. In making a critical examination of a text, it should be remembered that it is a means to an end, namely learning, and not a book intended for general enjoyment by someone knowledgable in the subject. For example, an elementary reader prepared for the first grade may contain such gems as "This is Spot. Spot is a dog. See Spot run." -- hardly material to cause excitement in critical literary circles, but still important in showing a child how the basic building blocks of the English language go together.

The major problem of the textbook writer is that of matching his style to the background of the student. (His problem is very much the same as that of the popular magazine writer.) The practice problems and examination questions included in the preceding chapters also affect the presentation of text material. The best way to write a text, of course, is to first teach the subject to a class, and then record in writing the teaching methods which experience has shown to work best.

The Workbook

Most textbooks include questions for the students to answer or problems for him to solve. A type of text which has become popular in recent years is the workbook. A workbook consists of a text, usually somewhat shorter than the normal conventional textbook, and blank pages on which the student makes his problem solutions. The first workbooks were on subjects in which the problems involved graphical constructions or drawings, and the problem sheets saved time for the student in providing the basic set-up for his solution. Standard worksheets also saved time for the instructor in correcting solutions by insuring that the right answers all looked alike, even to providing boxes for numerical results to be written.

The writer of this lesson used the workbook format for his textbook, Introduction to Engineering, which has equal numbers of pages for text and problems. He prepared an answerbook for instructors by filling in the correct solutions on a set of problem sheets. (Actually, the answerbook was done first, and then the solutions blanked out on the offset printing negatives for the problem sheets. By this procedure, it was guaranteed that graphical work for a solution would not run off the page, which would be embarassing for both the teacher and the author.)

The workbook possesses a curious economic advantage over conventional texts. With most textbooks, a certain number will be sold to students on the first year of an adoption by a college. On the second year, new sales may drop to 50% or less, because of the used books turned in by students for resale. The practice of many students of selling their old textbooks is poor, because a text which has been studied in detail becomes the best referencebook the student can have in later years. The workbook is likely to be retained because its resale value is nil, owing to the previous use of the worksheets. Each year, the sale is 100% new books. The students have the advantage of keeping their texts, with a file of solved problems, and the author profits by the lack of used books competing with new ones.

In writing a workbook, the author starts with a file of problems which illustrate the subject matter he desires to teach. The text is then written to show how the problems should be solved. Workbooks have a practical orientation; they do not expound complex theory, but tell how to solve real problems in the most efficient manner. Workbooks are popular in the study of such varied sub-jects as kinematics, mathematics, drafting, descriptive geometry, architecture, industrial design, and music. Also, the idea has been used in social studies, even when there are actually no pages on which solutions are written -- the name being used to suggest a kind of communal effort on the part of the class, with the instructor serving as a referee rather than a lecturer.

If you should find yourself given a teaching job in your company, you may be wise to start the assignment by preparing a workbook for the course. It will save time for you and the students, and insure a smoothly functioning operation.

Programmed Learning

Some years ago a new idea in learning was proposed, which has received the name "programmed learning." It makes use of a special kind of textbook, re-ferred to sometimes as a scrambled text. There has been a great deal of ex-periment and discussion with programmed learning, but very few of the special textbooks have been written. The reason is undoubtedly that the preparation of such a book is quite difficult, and even the enthusiasts on the programmed learning would rather spend their time giving learned papers on the subject at Miami Beach or some other convention center, than in the hard work of writing a scrambled book.

The textbook which is the foundation of programmed learning consists of hundreds of questions, each with multiple-choice answers. The student uses the book as follows.

On the first page he finds a general question about the subject, with multiple-choice answers. He chooses an answer, and then turns to a page number printed for that answer. On the page, he finds a comment relative to that answer. If his answer was correct, the student returns to page 1 and tries question number two. If his answer was wrong, the discussion so states (perhaps telling why), and gives a hint relative to what the right answer might be -- then refers back to the original question for another try. In making several tries on a question, the reader finds himself in a situation somewhat like that of a student working with a tutor, who asks questions and gives hints until he gets the correct an-swer.

Here is a very simple example of how Ohm's law might be taught in a pro-grammed book:

1. What does Ohm's Law state?

Answers: a. That electric heating is proportional to the current squared.

b. That current is equal to voltage divided by resistance.

c. That the capacity of a capacitor is proportional to the applied volt-age.

    d. That power is the product of voltage by current.

    e. That resistance depends on resistivity, length, and cross section-
al area.

For each answer there is given a page reference. For the foregoing answers, the comments might be as follows, each one found on the particular page it is referred to. _____

    a. Wrong answer. Electric heating, meaning the power in watts converted into heat, is equal to I^2R. This important equation is related to Ohm's law in-directly, but it is not Ohm's law. Ohm's law does not involve power or energy, but only the fundamental electric quantities of current, potential difference, and resistance. Try Question 1 again.

    b. Correct. Ohm's law, the very foundation of electric science, states that $I = E/R$. Now go on to Question 2.

    c. Wrong answer. Ohm's law has nothing specifically to do with capacitors, and furthermore, this answer is wrong physically. The capacitance of a capaci-tor is determined by the geometry of the capacitor -- its area and the thickness of the dielectric between plates, and by the dielectric properties of this insulat-ing material. Hint: Ohm's law has to do with the three fundamental electric quantities: current, potential difference, and resistance.

    d. Wrong answer. The statement is correct, Power = EI, but this is not Ohm's law. Ohm's law describes the fundamental relation between current, po-tential difference, and resistance.

    e. Wrong answer. This statement defines resistance, which is one of the fundamental electric quantities contained in Ohm's law. The others are poten-tial difference and current. Try again!

_____

If you have an interest in preparing a scrambled book as an exercise in logical writing, we suggest that you begin by finding yourself a willing victim, and try the questions on him verbally, writing down the questions and answers proposed, and keeping track of your comments on his choices.

Study of a scrambled book allows the individual student to learn at his own pace, and without wasting time by listening to the mistakes of others. If he knows an answer, he can proceed at once to the next question; if he doesn't, he is led by hints to the correct solution. The comments on wrong answers can al-so provide additional instruction by telling something about the incorrect state-ment. For example, in our illustration, Answer "a" is explained as an impor-tant law, even when it is not the proper response to the given question.

Training Manuals

We noted earlier in this Engineering Series that a "manual" or handbook is a

book kept "at hand" for needed information. Handbooks are really not instructional in nature, but are sources of information for persons already skilled in
the field covered. The name "manual" has also been applied to a teaching book,
used by those who want to learn the practical aspects of a vocational subject.
Training manuals are usually written to impart skills, with background knowledge a secondary object, and basic theory almost never covered.

There are thousands of manuals written for use by the Armed Forces, giving
training in the use of military equipment, and in the servicing and repair of that
equipment. Many of these manuals are prepared by the manufacturers of the
equipment under contract to the government. The terms of government contracts usually spell out in detail the format and style of a user's manual, or
service manual. Manuals which go over the dividing line, from information to
instruction, may be written much more freely. Even a strictly utilitarian operator's or maintenance manual becomes a better document when it is written
with a teaching viewpoint, and manuals offering this "extra" are appreciated by
the government agencies for whom prepared, and above all by their ultimate
readers.

A training manual is really a small textbook. It is usually short on complex
theory but long on practical applications of that theory. It may be divided into
chapters like a text, or into sections and sub-sections like an operator's manual. Unlike the latter, it may contain exercises, problems and questions for the
student to use to drive points home and check his comprehension of the subject.
It is usually written in a less formal style than a textbook.

Textbooks are usually published on speculation by the publisher, who seeks
school adoptions. A training manual, on the other hand, is most often published
by an organization for the use by its own employees, or those of its customers.
Textbook authors are paid traditionally by a royalty, which may be anything
from 5% to 15% of the selling price of the book. Writers of training manuals
are paid an hourly or monthly wage for their work, or sometimes a fixed sum,
under terms of a contract.

Training manuals offered at public sale include such paper-back books as
those on transistor applications published for transistor manufacturers, and
self-teaching books on such subjects as auto repair, animal husbandry, and
home ceramics. A highly specialized book of this kind is the do-it-yourself
"doctor book" which describes diseases and their symptoms, and tells about
simple home remedies and treatments. (Some doctors don't like these.) Books
on subjects ranging from flower arranging to oil painting seek the hobbyist
market, and particularly the ever-increasing senior-citizen population. Anyone
seeking a subject to write on, in the field of simple textbooks, can do much
worse than picking a hobby like surfing or photography to describe in an authoritative manner.

The greatest recreation of all, sex, is represented by hundreds of "training
manuals" of all degrees of propriety, from bland accounts of activities of bees
and flowers, to erotic paperbacks which use education as a feeble excuse.

Outlines and Cribs

We have mentioned the lecture notes mimeographed and sold privately by students, and the paperback syllabi published nationally in many fields. There are also printed one-page summaries of subjects like high school geometry, suitable for insertion into a ring binder, which students can use for last-minute cramming.

Another "literary" publication is the super-condensed outline of a classic piece of literature, to be memorized by students who just want to pass an English exam, and are too lazy or stupid to want to enjoy the rich recreation of reading a great novel like Huckleberry Finn. The writer of this lesson once perused some of these capsule classics, and was astounded to see the fascinating adventures condensed like laws of physics, into a totally lifeless catechism -- a dried up parody of living. While this sort of "instruction" must be called stultifying and wretched, we must note that human lazyness has created a good market for the enterprising writer who has read the real book. Perhaps you may be able to infuse some of the original spirit into a condensation of the original even to 1% percent of its original length.

Writing Correspondence Courses

We cannot conclude this lesson without a brief discussion of the correspondence course as a type of educational writing. As a correspondence student, you have been exposed to the format and style of our lessons, and have no doubt formed your own opinions as to the strong and weak points of instruction by mail. We shall now "take down our hair" and tell you some of our philosophy and approach to education carried out by correspondence.

The advantages of correspondence study are primarily ones of convenience. The student does not have to drive to classes meeting at arbitrary times and places. He is not tied down to the slow pace of a class which is geared to its most obtuse members or swept away by a class paced for more rapid learners. He can study at his own pace and put in as many hours per week as he desires, thus adapting the whole learning process to his own individuality.

On the negative side, the greatest problem is created by the lack of personal communication back and forth between student and teacher. Usually the best teaching is done face to face, and there is no real substitute for a classroom where one can ask questions, get answers, and be quizzed with other students supplying subtle encouragement and competition.

It is always a temptation to say, "Why go to school when the information is all in books at the library?" But when one tries to learn by reading an ordinary textbook, problems arise. Many questions aren't answered in the book. As a reader, you may not know how much or how little to study in one session; if you try to take in too much, you can become confused and discouraged to the point of giving up.

The correspondence course represents an attempt to present text material for home study in a form which minimizes the foregoing difficulties. By organization, format, and style of writing, a home study lesson seeks to create a classroom atmosphere through the printed page, and to anticipate and answer questions that normally arise. The examinations at the end of each lesson are not mainly tests by which the student is judged, but are learning devices and instigators of two-way communication between student and instructor. A correspondence lesson is an attempt to have your cake and eat it, and of necessity it can (for many people) only approach the effectiveness of good classroom teaching.

How does a correspondence lesson differ from a chapter in a textbook?

First, its length and content may be adjusted to that which can be studied by the average student in a certain number of hours. However, the course writer must subdivide the lesson into sections so that the student can break up his study into several reading periods. The lesson is thus a kind of mini-textbook, with short parts that can be digested in a half hour or so.

Each lesson will probably commence with an introductory paragraph, which outlines what is to be covered, and perhaps relates it to material presented in earlier lessons. Sometimes the introduction should justify the material -- that is, tell why it is important. (College students often grow tired of theoretical study whose practical application is not adequately pointed out by the professor.)

Each lesson may end with a Summary that calls to mind the main points covered in the lesson, reviews the conclusions and perhaps mentions what is to come in future lessons -- like the come-on in TV serials: "What will happen when John discovers that Cynthia has been unfaithful to him?"

The body of the lesson, between introduction and summary, develops the subject matter in an order which matches the normal learning process -- just like a textbook, a technical article, or even a technical report. New concepts should be compared to things already familiar to the reader. Examples should be given, using numbers when possible to create a quantitative impression of the matter.

Correspondence lessons should be written in a more informal style than textbooks. The reader can be addressed as "you" (instead of in the third person) in order to create a personal impression. The lesson writer may even go so far as to refer to himself as "I", e. g. , "With regard to transistor theory, I think it is sufficient for you to know that. .. " The writer of this lesson (that is, I) chooses to use third person, although he cannot give a convincing reason; it is just personal preference, based perhaps on the habit of writing regular textbooks.

Informal style relaxes the reader. In writing a correspondence lesson you must seek to visualize your reader and his situation when studying the lesson. Think of somebody who has put in eight hours on a job, and is tired to the point where it takes an effort to turn away from TV or bed, to study something that is best tackled first thing in the morning. This reader will enjoy an easy-going style, with points made simply, and perhaps even repeated in several ways with different comparisons. A certain lightness of style, and humor, usually also helps; you may have noted this approach in these lessons. Humor must be delicate and subtle, and the wrong kind can backfire. A serious student wants first of all to learn, and too flippant a style will annoy him as a waste of his time and money. What kind of humor and how much is right? The answer to this depends on the subject and the reader. Think back over the lessons in this course. Have some of the quips made you chuckle and relax, or did they annoy you as being out of place? You the student are the judge. If you liked a humorous comment, then this might be the kind to use when you are preparing training material for correspondence use. If you were repulsed, take that section as an example of what not to do.

A characteristic of some of our correspondence lessons is the use of practice problems and questions, with solutions given at the end of the lesson. Such questions replace class recitation; they give the student a chance to think out answers on his own, and then compare them to the right answer (or at least, to the teacher's opinion as to what is the right answer). The best place for practice questions is at the end of each main section of the lesson. By stopping reading and taking up a pencil, the student gets a good change of pace. The questions also give him a check on how well he understands the section, and they may send him back to re-read parts in order to understand a point which wasn't clear the first time around.

Each lesson concludes with a test which is mailed in for personal review by the staff of the school. The test questions should be similar to the practice questions and problems. Some correspondence schools do not use questions, or problems requiring the checking of numerical calculations. Such questions take so much time on the part of the school staff that instructor costs become excessive. In Grantham technical lessons, multiple-choice, true-false, essay, and show-your-work questions are all used. Many multiple-choice questions require numerical calculation. Generally, five answers are given to choose from. One of these must be correct and the others wrong. The wrong answers must usually be sufficiently different from the right so that slide-rule error will not make the difference. At the same time the wrong answers should not be entirely absurd. In selecting the wrong answers, the writer should seek answers that might result from common mistakes of concept, theory, or calculation. For example, if the question is to calculate the energy stored in a $3\,\mu F$ capacitor when 100 volts is applied, the correct answer is $\frac{1}{2}CE^2$ which is

$$\tfrac{1}{2}(3 \times 10^{-6})(10,000) = 1.5 \times 10^{-2} \quad \text{or} \quad 0.015 \text{ joules}$$

An expected mistake might be to substitute the number of microfarads instead of farads, which would give 15,000 joules. Another error might be to forget the $\frac{1}{2}$ factor, getting 0.03 joules. A combination of the two errors would give 30,000 joules. To find a fifth answer choice, one might just slightly move the decimal point in the right answer, e.g., 0.15 joules. A tricky answer might be to give the correct number but the wrong units, e. g. 0.015 coulombs.

An ever-present hazard is that the correspondence test writer may make an error in his own calculation, so that none of the printed answers is right. This can be embarrassing, but it most likely is not harmful to the student, as it may make him think very carefully before sending a sizzling comment to the school about their dumb questions. Your writer has frequently played dumb in lectures before a live class, making a mistake for the students to pounce upon, because this wakes everybody up; the game of catch teacher in a boner can be a fine way to make the students learn. In the case of a correspondence lesson, however, I believe that there should not be deliberate errors. They can create serious misconceptions, and waste the time of a student whose hours are valuable. So if you find an error in my tests, it is a real mistake and I will appreciate hear-hearing about it.

To my knowledge, the scrambled-book technique has not been used in a correspondence course. I have considered the possibility of using this technique, and have discussed it with Mr. Grantham. Perhaps we'll try it sometime. If you find yourself writing lessons for home study, and try the programmed scrambled-book approach, we'd appreciate hearing from you as to its success.

Home study today is a growing method in education, as everyone's time becomes more valuable. In spite of the competition of free night classes in high schools, more and more people are turning to the correspondence course as the best way to learn exactly what they need for professional advancement. Correspondence courses cover fields which are changing rapidly, so that the lessons need to be upgraded frequently. The need for technical writers who can revise lessons and prepare new lessons is a growing one. If a correspondence school has its headquarters near you, it might be worthwhile for you to visit it and propose working for them on a part-time basis. At the start, you might act as a test reader, in order to gain firsthand knowledge of student backgrounds and the sort of questions they need to have answered. Then, when you are familiar with the style of lessons used, you could take on the job of revising and updating lessons, and finally, that of writing new lessons.

Summary

This lesson has discussed the preparation of a wide variety of teaching materials. There are two prerequisites for an educational writer: knowledge of the subject, and ability to write in a style that leads to effective learning. When the material is advanced, the writer must have thorough knowledge. Even elementary material makes severe demands on the writer in seeking ways to introduce and explain new ideas.

The writing of educational material is a form of teaching. The best experience for a prospective writer in this field is actual teaching, because only this gives a true understanding of the ways in which the human mind works in the process of learning, and shows what analogies and comparisons work best in making a new idea plain.

Textbooks and other publications intended for classroom use with a live teacher present, can place their main emphasis on the factual content, relying on the human teacher to answer questions and meet the psychological blocks to learning which may develop in the student. Many training manuals, and above all, correspondence lessons, must stand on their own feet, and seek to accomplish the whole job of teaching without a supplemental question answerer. These types of educational writing are much more demanding on the ingenuity of the writer than is the preparation of a classroom text.

All technical writing is in a sense educational in nature, in that it seeks to convey information to the reader. Educational materials, in addition, seek to create viewpoints and attitudes in those who study them. In technical materials, including correspondence lessons, one of the main goals is to develop a professional attitude in the student. A professional attitude is compounded of many factors, including pride and ethical sense. Most important, perhaps, is a feeling of professional solidarity. The written text must assist the classroom teacher in making the students realize that they are gaining more than an organized body of knowledge and skill, but are also receiving a passport to an exclusive world of the professionally educated in a field, beside whom others are mere laymen.

Practice Problems

In this lesson we have left the practice problems until the end, rather than offering some at the end of each section. The main reason for this departure from the usual policy is to allow reading of the discussion of all the types of instructional writing. A comprehension of textbooks, manuals, and the other kinds of material must be of a unified nature before one can approach any of these on a practical level.

The following exercises in instructional writing are offered as challenges to your ingenuity -- to your ability to find examples and comparisons to put points over to the type of student specified. Obviously, an analogy that would appeal to a mature student might confuse a child; the experience and background of the student is most important in determining the method of presentation and explanation.

1. Describe electrical resistance on a physical basis to a 14-year old, who knows about Ohm's law but nothing about the properties of electrons.

2. Write a discussion of how the digits 0 thru 9 are combined in the decimal number system. Write for the high-school-senior age level, as regards comprehension and sense of humor, but put it in the style of an Aesop fable, supposedly written for very young children. Specifically, make it the story of a shepherd named Og who had too many sheep to count, until he visited the village mathematician who let him in on the secret of the decimal number system. If necessary, review just what a number system is, and then convert the dry mathematical explanation into a lively little parable.

3. Look up the biography of Nicola Tesla, inventor of the induction motor, and write up in a few hundred words, his dealings with Westinghouse in selling the patent of the motor. You'll have to do a bit of library research to find the best biography of Tesla.

MORE PRACTICE PROBLEMS FOR THOSE WHO ARE MORE AMBITIOUS

4. Write instructions for finding square root by the long division method, to instruct a junior high school student in the 13-14 age group. Get a volunteer student to read it, try it out, and ask questions on any points that are not clear. On the basis of his questions, rewrite your lesson to anticipate them, and try the new version on another volunteer. (We won't give you a "solution" on this one. You can find the information in some beginning algebra textbooks, but perhaps you can improve on the one you find. Your writing style here should be somewhat like that of a correspondence lesson.)

5. Describe how vacuum tubes work, for a high school senior.

6. Tell about integrated circuits, and how they are making a revolution in electronics.

7. Find out from a biography how Lee de Forest came to invent to the triode tube, and write this up in a light style for reading by mature adults.

8. Explain how transistors work, from the basis of solid state physics. Explain clearly what "holes" are, and use the shifting of cars in an auto parking lot as a comparison.

9. Write an explanation for solving quadratic equations by "completing the square," and derive the quadratic formula. Make this dry subject lively and interesting.

Discussion of Practice Problems

1. Electricity is made of electrons, which are tiny pieces of electric material which repel each other violently. They are much smaller than atoms, and free ones are able to sneak in between the atoms that make up matter. In metals, there are a lot of loose electrons which wander around among the atoms. Suppose we take a piece of copper wire and connect the ends to the plus and minus terminals of a battery. The minus terminal has an excess of electrons, which spill out into the wire, repelling the electrons already there. These in turn push other electrons farther along the wire, until finally the electrons at the other end are pushed right out of the wire into the positive terminal of the battery. A positive charge means that there aren't enough electrons, so really we can say that the positive terminal attracts these displaced electrons.

Now when the electrons flow along the wire, they constantly bump into the atoms. Every time a collision occurs, the electron is slowed down and the atom is made to vibrate. When atoms vibrate more, we say that the material is hotter, so the result of moving the electrons is to heat up the wire. Also, when the electrons are slowed down, the battery has to exert more push to get them going again. It turns out that the extra electrical energy to get the electrons moving again is exactly equal to the heat energy given to the wire.

We describe the whole effect by saying that the wire has resistance to the flow of electrons, just as a pipe containing cotton batting offers resistance to the flow of water thru it. We define the unit of resistance, the ohm, as that amount of resistance which requires one volt of potential to push one ampere of current through.

(The foregoing answer goes somewhat beyond the answer to the given question. The last paragraph above would constitute an answer to the question as stated, but without the build-up of the earlier paragraphs, the young student might end up very confused because he didn't have the physical concept of electrons threading themselves in among the atoms.)

(Other comparisons that might appeal to 14-year olds might be that of ants working their way through deep grass, bumping into the grass blades, or perhaps a troop of boy scouts finding their way through the woods in the dark, and bumping into trees. The latter analogy may appeal to the humor of a young reader, and will be more effective for this reason.)

2. Once upon a time there was a shepherd named Og. The grass was lush and Og was good at killing wolves, so the flock increased rapidly and Og became a very rich man. One day Og wondered just how rich he was, but the trouble was, he did not know how to count. Now this might make one think that Og was retarded, but this was not so; in those days, only the very wise men, such as mathematicians, could count. So Og went to the village and sought out the wisest man who, of course, was the local mathematician. Said Og, "I want to know how many sheep I have, but the trouble is, the sheep know how to multiply but I can't even count." So the mathematician thought for awhile (he really knew the answer right off, but being a consultant, he had to make it look difficult) and then he said, "Og, my boy, what you need are some digits. It so happens that I have a few spare ones left over from a special order that wasn't picked up, which I'll let you have at a special reduced price." So Og took the digits home and began to count sheep. First he had a zero, and this he threw into the ash can since it was absurd to count no sheep. Then he counted, 1, 2, 3, 4, 5, 6, 7, 8, 9. Here he ran out of digits, but he hadn't run out of sheep. Now Og was sorely distressed, and he went back to the mathematician and said, "How about some more of those bargain-priced digits?" And the mathematician said, "You're on the wrong track, Og my boy. It's a losing game to just get more digits. You can never keep up with the sheep, and if it was rabbits it would be even worse." So he whispered in Og's ear, and Og smiled and went back to the flock. First he went to the ash can and recovered the zero (fortunately, the trash pickup wasn't until the next day) and summoned the sheep. Then he counted, 1, 2, 3, 4, 5, 6, 7, 8 and 9. Then he put the zero down to show that he was starting to count again, and put a 1 in front of it (actually, a passable copy he had carved out) which gave him 10, meaning the first repetition of the original digits. Then he wrote 11, 12, 13, 14, 15, 16, 17, 18, 19, where all the 1's meant "first repetition." Following the 19, he wrote 20, 21, 22, 23 etc., for the second repetition, and 30, 31, 32, 33, 34 etc., for the third repetition, right up to 99. Then he put 100, meaning the first repetition of the repetitions, and started all over again.

The sheep were flabbergasted, because no matter how fast they multiplied, Og could always catch up with them. The reason was, as one wise old ram said to his favorite ewe, that Og wasn't playing fair. Instead of numbers, he was using a number system.

3. In the early days of the electrical industry, two giants battled for control of power distribution and manufacturing: Thomas Edison and George Westinghouse. Edison favored direct current systems, while Westinghouse had acquired the patents of a French inventor for alternating current generation and the transforming of voltage. Now there are many pros and cons for d. c. versus a. c., which make a very interesting story, but not for right now. Our tale concerns a very serious problem for Westinghouse in those days. A. c. was fine for long distance transmission, and it worked with electric lights just as well as d. c. But when it came to electric motors to drive pumps and fans and machinery, direct current had all the advantage, with the powerful shunt and series motors. Some kinds of series motors will also work with a. c., but these are no good for fans or pumps. So what Westinghouse desperately needed was a good reliable a. c. motor for his customers, before they all abandoned him in favor of d. c. and Edison.

Old George spent many sleepless nights, after board meetings of his company, wondering if it would even survive.

And then one bright morning, a young Italian walked into his office. His name was Nicola Tesla, and he had a simple business proposition to make to Mr. Westinghouse. Would Mr. Westinghouse be interested in a good reliable a. c. motor? After Mr. W. had recovered his composure, he replied, yes he would, and please sign right here, Mr. Tesla. The agreement offered was a very good one for Tesla, for it gave him one million dollars down, and one dollar for every horsepower of the Tesla induction motor built by Westinghouse. The million dollars was pretty dazzling, especially in those days, but what really made the deal out of this world in the long haul was that dollar per horsepower. As any schoolboy could figure, this would very quickly make Tesla the richest man in the United States, richer than Westinghouse and Edison put together, and probably with Rockerfeller thrown in (this was before Rockefeller had struck oil).

So the contract was signed with flourishes, but before the ink was dry, members of the Westinghouse board hired themselves a schoolboy who figured out that this agreement would make more for Tesla than for them. So the board called Mr. Westinghouse in and said, in effect, George old boy, we appreciate how the induction motor has saved our necks, but don't you think you went a little overboard on that dollar per horsepower bit? In fact, they said, our accountant here figures that unless we can break the contract this blasted inventor is going to make more than we are. This isn't right, because soulless corporations like us are supposed to exploit poor young inventors, and not the other way around. In fact, George, you'd better get cracking on this, or we may even be looking for a new president.

So Mr. W. called in Tesla, and having taken the measure of his man who was young, idealistic and very gullible, he said as follows: "Nicky my son, you are a very great inventor with a brilliant future. This induction motor is only a start, and with this great fortune you have you will be able to astound the world and make ten million on the next invention. But would you feel right if you knew that through an unfortunate error on our part, we -- your great benefactors -- might go broke and lose our homes and our shirts? You don't really need that dollar per horsepower, and wouldn't it bother you deep inside to think that every one of those dollars was hurting us?"

We don't really know if George and Nicky actually broke down and wept together, but we do know that Tesla tore up the contract and went away with just his million dollars, head high and facing the world with a clear conscience. As to Mr. W., we can't say, because soulless corporations don't have consciences. The only thing wrong with Tesla's great gesture was that he never made another invention worth even one million dollars. He invented the Tesla coil, which students like to build for school open houses and science shows, and he fooled around with the transmission of power by radio, which didn't work at all. He spent his million dollars, and some more millions invested by people who believed what Westinghouse had said. Towards the end of his life he became interested in communicating with spirits by radio, and when he died in the 1940's, he was penniless and living in a cold walk-up flat in New York City.

Now we aren't sure what the moral is to this story, except that perhaps it proved that P. T. Barnum was right when he said (of suckers), "There's one born every minute" (and he wasn't thinking of fish). We don't want to be unfair to Mr. Westinghouse by making him out a villain, because he wasn't. He just got carried away, and if he had only made it 10¢ per horsepower, Tesla would have been only moderately rich, the Westinghouse board would have grumbled a bit and then been satisfied, and -- who knows -- Tesla might have been successful in reaching the spirit world.

In the foregoing, a very light style has been used, to make a kind of grown-up bedtime story of what could easily be a very dry account of a business transaction. There are tricks in this style of writing. For example, the expression "good reliable a. c. motor" was used in two places, as a kind of parody on the way business men talk in their ponderous fashion. The late Damon Runyon used such heavy-handed phrases in his Bowery dialog to create the images of his outlandish characters.

Another way to tell the above story would be to put the imaginary conversations between Westinghouse, Tesla, and the Board in play form. The main danger in writing in a light style is that of making it too flippant, so that the humor annoys rather than amuses. Generally, it is best to employ such tricks sparingly -- like curry powder, too much can spoil the whole dish.

Space does not allow full discussions of possible solutions to the other writing problems presented for your practice. We urge you to spend as much time as you can on these, and then to try them out on your ultimate audience and most severe critics -- people who want to learn from your writings. No matter what the style of your piece, if it results in learning it has been successful.

TEST TC-8

TRUE-FALSE QUESTIONS

1. It helps to teach a course before writing a textbook on it. . . . . _____

2. A syllabus is similar to a correspondence lesson in being intended for study without the aid of a teacher. . . . . _____

3. A syllabus is the next step up in expanding a course outline. . . . . _____

4. Handbooks should be written so that they give a good picture of underlying theory. . . . . _____

5. The imparting of factual knowledge can be done only by a live teacher. . . . . _____

6. A textbook should also be a compendium. . . . . _____

7. Most correspondence courses make use of programmed books. . . . . _____

8. The workbook as opposed to the textbook offers certain financial benefits to the author. . . . . _____

9. Training manuals usually do not give very much theory on their subject. . . . _____

10. An authoritative compendium is likely to be more formal in its style than a correspondence lesson. . . . . _____

ESSAY QUESTIONS

Write a 300 word (approx.) instructional essay covering one of the following topics:

1. An explanation of how a vidicon camera works, written for an intelligent mature layman.

2. An explanation of the Xerox process of document copying for a new employee of the Xerox company.

3. Describe how a flip-flop works as part of a general course for technicians.

4. Write, as part of a correspondence lesson, a discussion of integrated circuits.

5. Describe the subtractive process of color photography, for a text for photographic amateurs.

LESSON TC-9
Editing Technical Materials

Introduction

The work of the technical writer is gathering information and then organizing and writing results in a manuscript for a publication. Between the typewritten draft and the completed publication there is a great deal of work which is generally called <u>editing</u>, but which might more correctly be called preparation for publication. It involves corrections of various kinds in the manuscript; arrangement of the material into chapters, paragraphs, and sub-sections; numbering of illustrations, charts, graphs, and equations; selection of page size and type style; and coordination among illustrators, artists, photographers, typists, typesetters, and printers. In material written under a government contract, the <u>editor</u> must also serve as an authority on the voluminous government specifications as to style and format, both to guide the writer and to ensure that the final product will be acceptable. Very few technical publications prepared under government contract are rejected on grounds of technical error, because most government editors do not have sufficient technical competence to find such errors! Many publications have been returned on grounds of non-conformity with editorial specs (short for specifications), by government editors who are happy to be able to discharge their duties in non-technical matters.

The motivation of technical editors is generally different from that of writers. Editors are not necessarily creative in the excitable prima-donna sense, and should be calmer and more reflective than writers. Obviously they must enjoy reading, for 75% or more of their time is spent in reading manuscripts. The editor's life is more sedentary; he stays at his desk instead of visiting others for information-gathering interviews.

A good technical editor must love painstaking, detailed work. Like an accountant, he must love to make every detail right, and produce a tidy job in which everything is in its proper place. When he feels that something is not right, he must want to go after the difficulty with dogged determination until he feels satisfied that it is correct. There is no place in technical writing for one who sweeps problems under the carpet, or hopes that troubles will go away when they are ignored enough. There are places where a perfectionist can cause costly delay and where compromise is needed to get the job done, but the technical editor's desk is not that place.

The conscientious editor may sometimes rewrite sections of an inept or badly expressed manuscript, but here at last there are limits. The editor cannot be a rewrite man; if the writer has poor style, the publication must ultimately go to press in that style. An editor can fashion crutches for a literary cripple, but he cannot make him into an athlete.

The editor's job can be subdivided into two general activities. These are first, the work of correcting style and putting the manuscript into the required format, and second, the general supervision of all activities which must precede publication. Let us consider first the work on the manuscript.

Copy Editing

The job of the copy editor is to make sure that the manuscript is in clear and acceptable English, and that it meets the style and format requirements of the company and the specifications of the customer -- the latter usually meaning governmental specs.

A copy editor functions somewhat like an English composition teacher. Mistakes in spelling, punctuation, and grammer must be found and corrected, and badly written sentences repaired when possible. It helps if the copy editor understands enough of the technology to follow the meaning of the discussion. Without some degree of understanding it becomes difficult to judge style, and an editor could even introduce technical errors by rephrasing what might appear to be an awkward wording. The copy editor, however, is not responsible for the technical accuracy of the manuscript; this responsibility lies with the writer, and the engineers or others from whom he has obtained his information and who must review the manuscript before it goes into editing.

The basic requirement of a copy editor is that he have a good vocabulary and an effective command of English. He must catch errors in spelling and punctuation. He must see to it that there is agreement in number between nouns and verbs (e. g., not one out of the large group is faulty, instead of "are faulty"), and that all of the other common grammatical slips that even intelligent educated people make do not creep into the report, proposal, or other technical document.

Along with a nose for error, the copy editor must also be realistic. He cannot correct every mistake, or clean up every piece of muddy prose. Such nit-picking would delay completion of the publication, and would hurt far more than the errors would. So of necessity the editor must sometimes let errors stand, not from ignorance but from the realization that time is money. He must know when to require rewriting, and when to reluctantly approve mediocre writing.

One of the prime functions of a copy editor is pruning. Technical writers, enthusiastic with a newly-mastered subject, tend to become verbose and repetitive. In their anxiety to be sure that the point is made unmistakably, they may make it over and over. Redundancy has its place in teaching, but publication costs are too great to allow it in technical reports (and busy report readers are apt to skip repetitions anyway). The editor with a sharp blue pencil and a sharper eye for unnecessary words can save his company thousands of dollars in publication costs.

One of the chief jobs of a copy editor for manuscripts which are part of a government contract is to know government writing specifications. These rules are detailed and are intended to be followed. They describe the format for tables, graphs, and charts; the style of writing captions for illustrations and the method of numbering and identifying them; and even the manner in which the writer must refer to an illustration in the body of the text. Also in the specs, terms that may be abbreviated are listed with the approved abbreviation. The editor must provide himself with a current copy of the specifications and be familiar with them, so that he can spot noncomformities in the course of reading without frequent reference to the specs themselves.

The following excerpts from an Air Force writing specifications show the nature of some of the instructions to writers and editors:

MATERIAL REFERENCES. All materials referred to, such as lubricants, sealing compounds, abrasives, etc. , shall be identified by Government Specification number wherever applicable.

NUMBERING PROCEDURAL STEPS. Steps of procedure, such as steps of assembly or disassembly, shall be assigned consecutive lower-case letters of the alphabet. The same procedure shall be followed in lists of equipment or other similar lists forming a part of the text.

NUMBERING PARAGRAPHS. Paragraphs shall be numbered consecutively within each section, using Arabic numerals separated by a dash. The number preceding the dash shall indicate the section (converted from Roman to Arabic) and the number following the dash shall indicate the number of the paragraph within the section. For example, "2-15" would be the fifteenth paragraph of Section II. Paragraphs in appendices shall be numbered in the same manner except that a capital A shall follow the index number.

The writing specs give precise rules for almost every situation that could arise in the preparation of a manuscript. Separate specifications have been prepared by the Army, Navy, and Air Force, and the use of the wrong textbook could cause trouble for a company holding a contract with one of these Services.

It is the responsibility of the writer to follow the appropriate specification, but it is also the responsibility of the editor to see to it that he does. Thus, ultimately the editor may get most of the blame if a manual is rejected for noncomformity with a spec. The situation is a little like the relationship existing between a workman and an inspector. It is the workman's job to make parts according to the blueprint, and he is responsible for what he does. It is, however, ultimately the inspector's responsibility to verify that the part is correct, and when he has placed his stamp of approval on the part, any mistake becomes his responsibility.

Production Editor

When a manuscript has been corrected and brought into conformity with the writing style required, it then goes into production. Production is the process of conversion from a typescript to a finished report, book, or other formal document. More than one copy is required -- and the number may be up into the hundreds or thousands -- so some form of reproduction is required.

In a company having a publications department, a production editor is assigned to the job of nursing a manuscript through the many activities required. The production editor is a combined expediter, decision maker, and coordinator who keeps the project from bogging down, and sees it through to the great day when a stack of identical copies of the document is placed on his desk.

The production editor is not concerned with the quality of writing in the manuscript, nor in its adherence to any writing specification. Obviously, if he notes a misspelled word he will bring it to the attention of the copy editor, but such things are not his responsibility. He probably will not even read the manuscript with any intent to understand its meaning. In a large company, the production editor may not ever see or speak to the writer, just as the shop foreman in a factory has no need for personal meetings with the draftsman who made the tracing for a blueprint his workmen use.

The personal requirements for a production editor are quite different from those of a good copy editor. The production editor keeps track of a factory -- a factory for the production of printed material. He must keep track of thousands of typed pages, charts, graphs, drawings, and glossy photographs, which somehow must end up together in proper order. His nightmares are concerned with such things as getting the wrong caption on an illustration, or a complex diagram placed upside down.

Among the basic jobs of the production editor is the determination of the type size and style, page size, line lengths, and type for headings, titlepage, and captions. He does not make such decisions afresh for each publication, since usually a definite company format has been established, but he must be sure that no problems will arise with the sizes he picks out. One of his first jobs is to estimate how many printed pages will result from the typewritten manuscript plus all the illustrative material. When the final publication is printed on a large press which prints four or more pages simultaneously, he must attempt to make efficient use of the paper stock, so that there are not too many blank pages at the beginning or end of the book.

Many types of publication, particularly in small companies, do not require the services of a production editor. These documents include reports having small distribution, where copies are prepared from typescript by direct offset or with a Xerox copier, and proposals for which only 4 or 5 copies are needed.

In small companies, the dual editorial functions may be combined in one person, or there may be varying degrees of overlap. For example, proof reading of printer's galleys might be done by the copy editor who is most familiar with the wording of the manuscript, or by the production editor who brings a fresh viewpoint less likely to overlook errors.

It should be pointed out that the duties of editors (copy or production) in a company are quite different from those of the man with the same title in a book publishing company, or on a magazine. Magazine and newspaper editors are personages of great power and influence, who determine content and even policy of their publications, and are responsible to no one but the owner or publisher. Newspaper editors generally grow with their papers, sometimes through several generations. Typically, a newspaper started as a kind of by-product of a small-town printing business, which mainly served to keep the press busy and also bring the printer to the notice of his neighbors. A typical news item would report a chruch social or picnic, and the extent of the coverage was usually closely related to the size of the printing order which the church gave the printer for programs. When the paper prospered and grew in circulation, the editor also grew into a person of importance in the community, as the means for filling one of man's greatest needs -- recognition. The pinnacle of this sort of editorship was reached by the late William Randolph Hearst, who was about as far from our copy and production editors as it is possible to get.

In technical book publishing, in addition to copy and production editors there is usually a "technical editor" who is responsible for the technical accuracy of the book.

Proofreading

One chore which is an inseparable part of publication is proofreading. Proofreading must be done every time copy is transcribed by human hands, whether by a typist or by a typesetter or a linotype operator. A philosopher once said, "To err is human, to forgiven, divine." Since readers of technical publications are not noted for their divinity or angelic qualities, one of the major concerns of a technical publications group is to minimize error!

In the old days of printing, type was handset from typecases. The typesetter built up the copy using an elongated metal tray called a galley. In more modern typesetting, each complete line is set by machine and cast in lead. The individual lines are then placed in the galley. With the type (either handset or machine set) properly lined up in the galley, it is then inked, a strip of paper carefully laid on top, and a hard roller run over the paper, causing a copy of the type in the galley to print on the paper. The result is a galley proof, which is read by as many people as possible for correction. The proofreaders may include the author, his wife, the editor, and almost anyone else who is literate. Through the years a collection of shorthand notations came to be recognized by printers; these notation symbols are today known as proofreader marks. By means of these useful signs, the proofreader can talk to the printer without a word being said. An editor who is not fluent in their use is considered by any printer to be illiterate and not worthy of respect.

Here are the marks, with their explanations in plain English.

⊙   Period   ∧   caret -- indicating where something is to be inserted

⸴   comma   ⸵   semicolon

⁚   colon   #   insert space

ꙩ	apostrophe
?	interrogation point
ꙩ ꙩ	quotation marks
☐	indent one em (space equal to that required for the letter m)
②	indent 2 em's
$\frac{1}{m}$	one em dash
$\frac{1}{n}$	one en dash (space of letter n)
ꙇ	push down lead or space
⌒	close up
V	less space
‖	align type
?	special query to the author (write it on the margin)
⌐ ⌙	symbol to indicate words or letters to be transposed. Also write "tr" on margin of proof
stet	"let it stand" (written when something has been accidentally crossed out). Also dots may be placed under the erroneously deleted matter, but the word "stet" should be put in the margin to make it plain to all printers.
spell	spell out (e. g. , six, not 6)
⑧	circle around figure means spell out
ld in >	insert lead between lines
[	move to left
]	move to right
⌐	move up
⌙	move down
( )	add parentheses

[ ]	add brackets
tr	transpose
out, s.c.	take out, see copy
=	straighten line
ℯ	delete -- take out
X	symbol used in margin for broken letter (defective type)
¶	make paragraph here
no ¶	do not make new paragraph here
wf	wrong type font
ⱽ⫫	equalize spacing
≡ (or caps)	use capitals
≡ (or sm. caps)	use small capitals
lc	use lower case instead of caps
ⱽ⁸ ₈	supercripts or subscripts
_____	use italics
rom	use Roman type (not italics)
bf	use boldface type

Reproduction Processes

All technical writing (except perhaps the writing of an informal memo or progress report) involves some kind of multiple reproduction. A knowledge of the various processes is useful to the writer, and necessary for the production editor in any graphics and writing division of a company. Draftsmen, artists, and illustrators also need to know what can and cannot be reproduced, and what are the limitations of the various methods of reproduction.

Let us consider the common reproduction processes, as they apply to manuscripts, line drawings (such as diagrams, charts, etc.), and photographs, starting with those which are best adapted to short runs, and going up the scale to long-run processes.

PHOTOGRAPHIC REPRODUCTION

A simple and obvious method of reproduction is photographic: make a photo negative of each page in order, and print the required number of copies either by contact printing or enlargement. This process is rarely used because of the high cost of materials, particularly of bromide photographic paper, and the relatively high labor costs. If a few copies need to be made from an original available as microfilm negative, they may be printed by enlargement, but even in these cases, it is usually better to make a full-sized master copy, which is then duplicated by some other process. An exception is made in the use of glossy prints, either in black and white or in color, which are hand pasted into reports printed by other methods. Such an illustration, appearing as a front piece to a report or proposal, may be well worth its cost in showing a product clearly as well as enhacing the general appearance of the publication. A century ago, lithographs were sometimes inserted in this manner into printed books; today of course fine halftone engravings can be printed along with the text and linework copy.

REFLEX-DIFFUSION PROCESS

A variation on photographic copying, which does not require the use of a camera, is the so-called reflex process. In this process, the sensitized negative paper is placed face to face with the master copy, and the two are placed over a uniform light source with the negative paper downward. The light thus must pass through the sensitized paper, be reflected from the master copy, and pass again into the negative. The photographic characteristic of the negative is as shown in Fig. 1, which is a plot of exposure against photographic image den-

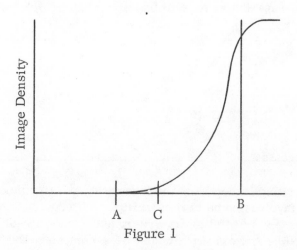

Figure 1

sity. The response of the paper (i. e. , darkening) is slight at first. At a certain critical exposure, the characteristic curve becomes steeper so that a relatively small additional exposure produces complete darkening. In use, the light source is adjusted so that the exposure produced when the light passes through the first time corresponds to point A. If the master copy is white, the reflected light adds an increment of exposure AB, which is sufficient to cause complete exposure.

If reflection is from a black line or part of a letter, the exposure is much less -- e. g. , AC, which produces no darkening. By proper adjustment of brightness and exposure time, the "knee" of the curve, where the bend occurs, can be placed so that the initial light plus the small amount reflected by a dark part of the copy will produce essentially no exposure. The somewhat greater exposure obtained by a reflection of perhaps 70% will create a dense image.

In copying, the reflex principle is usually combined with the diffusion transfer method of producing a positive image from the paper negative. After development, the negative is run through rubber rollers in contact with the positive paper, and the coating on the paper turns black on contact with unexposed portions of the negative. The sandwich comes out of the rolls "damp-dry" and the positives are ready for use in five or ten minutes. The negative copy is usually discarded, but if it is wanted for any reason, the image can be made permanent by immersion in ordinary photographic fixer or hype for a few minutes.

The foregoing process is capable of very high quality reproduction. The writer of this lesson has used it many times to copy illustrations from various sources. The copies were then pasted up in any desired arrangement for photo-offset reproduction. An example is found in the montage of sketches and symbols forming the decorative outside cover of some of the Grantham lesson books. In this montage, the slide rule scales, for example, were taken directly from an actual slide rule.

The cost of the above process is about 15¢ per $8\frac{1}{2}$x11 page, including the chemical solution. Each sheet can be processed in about 30 seconds. When this time is combined with the time to place the sandwiches of master copy and negative in the exposure box, the total becomes prohibitive for the reproduction of more than six or eight copies of a document. The process will copy fairly coarse halftone dots, but results are poor when it is used with fine-screen halftones, or original photographic prints.

DRY ELECTROSTATIC COPIES

The bugbear of all photographic and diffusion transfer processes lies in the use of wet solutions and the need to dry out the copies. Various dry processes have been devised, including the "Thermofax" system using a special paper which is darkened by heat. The first real success in dry copying came with the Xerox process.

The Xerox process is subtle in theory, and sounds so complicated that one must marvel at the genius of its inventor, Carlson, and the engineers who have so successfully automated the machine.

In the Xerox process, an electrostatic charge is applied in the dark to a sheet coated with metallic selenium. Selenium has a very high resistivity in the dark, but when light shines on it, photoelectrons are released which make it electrically conductive. As long as the sheet of selenium is in the dark, the surface electric charge remains, but if light shines on any part, the charge on that part at once leaks off to the backing plate. The image may be a page with typing or black linework; when the image (in light) is thrown on the selenium sheet, the charge remains only along the dark lines of the image.

Following exposure, a fine black dust is sifted over the surface of the selenium sheet. The dust acquires an induced charge and adheres to the selenium along the charged lines, but slides off where the selenium is electrically neutral.

The selenium sheet (mounted on a drum) is next brought into contact with a charged sheet of ordinary paper. The black dust is transferred by electrostatic attraction to the paper. The paper is then quickly heated, melting the powder which fuses into the paper making the image permanent on the paper.

Following the original Xerox invention, many other companies produced dry copiers utilizing somewhat similar processes, but most of them use special sensitized paper. Any ordinary paper may be used in the Xerox system.

The cost of Xerox and similar copies is about 5¢ per sheet. The lack of chemicals and the short time needed to make a dry copy has made Xerox preeminent in the field of short-run copying. The constant effort of the company is to increase speed (and to a lesser extent, decrease cost), in order to make Xerox complete with offset printing for short runs. When all cost factors are considered, including the pay of an operator, Xerox is cheaper than printing for runs of less than 50 copies and in modern machines is competitive for hundreds.

Xerox does not produce as sharp a reproduction as the wet chemical process mentioned. When closely examined, the lines are seen to be somewhat gradular due to the size of the grains used. Also, large black areas do not reproduce black all over. Because of electrostatic shielding effects, the grains tend to adhere only at the edges of such areas, leaving the insides white. For this reason, Xerox is poor for many kinds of linework, and unacceptable for the reproduction of continuous-tone photographs.

SPIRIT DUPLICATORS

A process widely used for runs of less than 100 copies is known commonly as the Ditto process, from one of the proprietary trade names. In this process a master copy is prepared by typing or drawing on a sheet of special paper which has another sheet against the back that is heavily coated with a special carbon ink. A thick reversed image of the typed or drawn copy adheres to the back of the sheet. The master is then placed on a drum, with the carbon-inked side out. This master is slightly dampened with an alcohol-water mix and then rolled against paper. After about 100 copies the ink is exhausted and the master sheet must be discarded.

The spirit duplicator has a unique advantage shared by no other reproduction process: more than one color can be printed at one time. After the original image has been applied (usually in a kind of purple ink) other carbon sheets can be slipped under, and additional lines or typing made in red, green, blue, or black. The red and green are bright clear colors, but the others are rather muddy and produce fewer good impressions. When the master sheet is run through the printer, all colors come out simultaneously and, of course, in perfect register.

The great drawback in the use of the ditto process lies in the limited number of copies than can be printed from one master. Various attempts have been made to duplicate masters by means other than retyping and redrawing. One successful copying process places a master copy on a drum which is scanned by a photoelectric cell. A second drum on the same shaft has a master-carbon sandwich, with a vibrating stylus which strikes it every time a dark line passes under the scanner. A fair duplicate of the original material is produced on the ditto master in about 15 minutes. By this method an unlimited number of ditto copies may be printed.

Ditto copies are quite cheap. To the cost of the paper there must be added only the price of a master set (about 15¢, or 0.015¢ per copy) and a small amount for the alcohol mixture. The printed result is relatively crude, so that spirit duplication is usually confined to in-house reports, memos, etc. Only a very small and poor company would put out a proposal in ditto.

MIMEOGRAPH

Under this trade name we describe the stencil duplicators. The use of stencils is very old, and indeed the mimeograph machine was the first truly successful office duplicator. This machine uses a stencil consisting of a porous paper sheet dipped in gelatine so that it becomes impervious to printing ink. When lines are typed or drawn on the stencil, the gelatine is pressed aside, leaving a porous web which supports the centers of closed letters like "o". The stencil is then wrapped about a drum with an ink-soaked pad underneath, and rolled against the paper stock. Ink oozes through the letters in the right amount to print.

Mimeograph stencils can be used for thousands of copies, becoming useless only when the webs break so that the centers fall out of the o's. If a stencil is well cleaned and blotted, it can be put away for reuse. The stencils are cheap, as is the paper (which is rather soft and porous), and the main drawback in the process is the fact that it is hard to keep the messy ink off the operator's hands.

Stencils can also be reproduced by the process described for ditto masters, and the possibility of making stencils from previously printed material has greatly increased the flexability of the stencil process.

OFFSET PRINTING

The most widely used printing process for technical documents is offset printing, also called lithography.

Lithography, a term meaning printing from stone, has been in use for many centuries. Originally, it was done with the aid of a type of porous stone, having one surface smoothed and flat. The material to be printed was drawn on the surface of the stone with a greasy crayon, in reverse. The stone was then wet, but the water did not adhere to the grease. A greasy ink was next rolled over the surface, but it was repelled by the wet stone, adhering only to the crayon lines. A sheet of paper was then placed on the stone and pressed against it so that it took up the inked image.

When all the prints desired had been made, the stone could be resurfaced for another picture. Lithographs had the disadvantage of printing in reverse, which required that the artist make a mirror image of his final subject.

Toward the end of the 1800's, the idea of printing from the stone onto a rubber blanket on a large roller, and then transferring the image to paper, was developed. This was <u>offset</u> lithography, in which the image was transferred or offset temporarily to the rubber before final printing. The advantage, of course, was that the image on the stone was identical to that on paper, avoiding the mirror-image problem.

More recently it was discovered that other surfaces than stone could be made receptive to both water and oil, so that the wet-oily contrast could be used for printing. Specially prepared aluminum and paper sheets have been developed to receive images typed with an oily ribbon, or drawn with a special greasy pencil. Actually, ordinary typewriter ribbons, either fabric or plastic carbon coated will work quite well for typing, and any soft graphite pencil will make acceptable offset plates.

Offset plates may also be made photographically on sensitized aluminum plates, printing by contact through a high-contrast negative. The sensitivity of the plate is quite low (an exposure of around 45 seconds is needed in bright sunlight) but the photographic grain is exceedingly small.

Photo-offset plates made today are usually either aluminum or paper. The cheaper paper plates break down after a hundred or so copies. Better grade paper plates are good for as many as 1000 copies or more, while metal plates can print many thousands of copies before the printing surface begins to detach from the base metal.

Offset printing plates may be made by direct exposure from the master copy (for example, by reflex copying) or by "burning in" the image from a high-contrast photographic negative. The standard procedure is to "burn-in" the image.

PREPARATION OF THE IMAGE

The original copy may be prepared in any of several ways. For the highest professional quality, the type should be set in metal, and photographed from a printed proof sheet made on very smooth stock in an accurate proof press.

However, most photo-offset master copy is typed, using a carbon-coated plastic ribbon for a sharp impression, or produced by a relatively new process called "photo-typesetting."

Most ordinary office typewriters give the same space for each letter of the alphabet -- 1/10 inch for pica type and 1/12 inch for elite style. To give a uniform appearance to words, the broad letters like m and w are compressed, while the narrow letters like i and j have expanded serifs to make them fill the space. In printing from movable type, the letters have various widths, as determined by type designers, which gives the pleasing professional appearance we associate with high quality printing. This print-like quality is approximated in the IBM Executive typewriter by providing several widths of letter, expressed in units of 1/32 inch. Narrow letters, as i and l, are alloted 2 units each. Most lower case letters, such as a, b, c, n, etc, have 3 units each. Lower case w has 4, and m has five. Most capital letters have 4 units. The numbers all are of 3 units, to allow for tabulation. Number 1 is separate from lower case el (ℓ) (which doubles for the "one" in conventional typewriters) since the "el" has 2 units, and the "one" must have 3 to conform to the other figures.

The page you are reading was typed on one of these IBM Executive Typewriters. The name of the IBM type face used here is HERITAGE

A newer development of the IBM company in typesetting machines for offset printing is the "Composer." The Composer is, like the Executive, essentially a typewriter, in that the images are placed on the paper by what is called the "strike on" process. The Composer provides a greater number of letter widths than the Executive typewriter, so the type produced looks more like the traditional type that has been used in books and other publications for many decades.

Since the Lintotype and Intertype machines used to set metal type work with melted lead which cools and hardens to produce each separate line of type, a job produced on these machines is referred to as "hot type." That is, a job set in hot type is one in which the type used to print is metal (actually lead). On the other hand, any typing process in which the letters are typed on paper for photographing is called "cold type." During the 1960's the printing industry largely converted from hot type to cold type. However, some cold type was used long before 1960, and some hot type is still used today. As mentioned earlier, for high quality printing hot type is often used to make the photographic negative for offset printing purposes.

A requirement of professional appearance in printed matter is that the right-hand margin be straight, and not jagged as it comes out in ordinary typewriting. Such a straight margin is said to be "justified." In setting hot type, the justifying is accomplished by inserting spacers as needed between words after the line is set but before the hot lead is poured. In typewritten copy, justification is most often done by typing each line twice. On the first typing, the line is stopped a little short of the desired length, and the difference (usually in "units" as defined above) is noted. During the second typing, these extra units are inserted by adding a few to each space between words, which will stretch the line to the required length. In special coldtype composing machines, the additional space is added by an electronically controlled system that eliminates the need for the second handtyping; the second typing is automatic.

As an example of how extra spaces are used to justify a line, consider this line:

The power supply consists of a bridge rectifier connected to...

The three periods of 3 units each were typed to bring the line out to the desired righthand margin. In order to put the nine additional units into the line, the typist puts 3-unit spaces between words, instead of the 2-unit spaces used in the initial typing. In this case, the spaces following the first eight words are increased from two to three units each to bring the line out, and the effect is:

The power supply consists of a bridge rectifier connected to

Only the last space, between the words "connected" and "to", still has two units. Can you see the difference between the first and second typings?

After offset printing had developed sufficiently to create the demand, and electronics had developed to the point that "electronic word processing" became economically practical, a typesetting revolution got underway in full force in the early 1970's. The process at the core of this revolution was neither that of hot type nor that of cold type, but was a newly developed technology called photo-typesetting. In this process, a keyboard, much like that of a typewriter, is used by an operator to code-perforate a paper tape (or code-record on a magnetic tape) with the information -- letters, numbers, and symbols to be set, along with instructions for type face and size, line length, etc. -- which a photo-type-setting machine will use later to produce the type at up to approximately 1000 words per minute.

The process in the photo-typesetting machine is electronic and photographic. The various type characters are transparent areas in rotatable discs. A light beam shining through the selected characters exposes a photographically sensitive paper sheet, one character at a time, to produce the type-character images. An electronic computer, guided by the input tape, directs the rotating discs and light beam, to the end of putting the right characters photographically at the right places on the copy sheet.

The copy comes out of the machine in the form of a long continuous sheet (or roll) of paper. Then, this type is used in the same manner as any other type-on-paper, for "paste up" in preparing the pages that will be photographed to make offset-printing negatives.

The negatives for words, line drawings, and halftones are stripped (taped) together on a sheet of semi-opaque paper in the proper position for the desired pages, and the paper is cut out around and removed from the parts of each negative having copy which is to be printed. The negatives must also be "spotted." That is, unwanted translucent places (caused by bits of dirt on the copy) are painted over with opaque.

Good offset printing is a fine art, requiring special skills on the part of photographers, printers, and other technicians. The production editor cannot be expected to be proficient in all of these jobs, but if he understands the processes and their limitations, he will be able to help create a high quality product at minimum cost.

Printing from Type

The highest quality publications are set originally in hot type. The actual printing may be either on a letterpress (directly from the metal type), or on an offset press which uses plates made photographically from printed proof sheets.

The great invention in printing was that of movable type, which was made by Johann Gutenberg of Mainz, Germany, in 1440. Gutenberg formed his letters in a style based on German manuscript lettering, which is called Black Letter. Italian type designers developed the basic Roman style, and the slanting type called Italic which is sometimes referred to as Cursive because it is somewhat imitative of script. In all of the foregoing types, short lines were placed across the terminations of the letters; these lines are serifs. The serifs were used by writers of script manuscripts to cover up the otherwise jagged endings of pen or brush strokes. A type style which eliminates the serifs is the Gothic, or "sans serif." (Sans is a French word meaning "without.") The history of type styles forms a most interesting chapter in any book on printing, and a study of this subject is both pleasant and valuable in a professional way to a production editor, as well as to other editors.

By the middle of the nineteenth century, printers began to seek mechanical ways to substitute for hand-setting of type. Many schemes were tried, but the most successful was the Linotype machine, invented by Otto Mergenthaler. In this machine a keyboard assembles, not a line of movable type, but one of brass moulds. Hot type-metal (lead) was then forced against the moulds, making (upon cooling and thus becoming solid) a single piece of metal which would print a whole line. Such type was easy to handle, and especially to space out by inserting thin strips of metal between lines. There was the disadvantage that a mistake required recasting the whole line, but this was easily done.

There is an interesting story told concerning the patents on the Linotype. Patents generally require several years between first application and final granting. During this time there is a rather slow correspondence between the inventor and the patent office, with six months allowed for the former and unlimited time for the latter. The life of the patent is 17 years after it is granted. Now one might think that an inventor would want to get his patent issued as quickly as possible. There is an advantage, however, to delay. The inventor is really protected from the date of application, so that his monopoly is 17 years plus the time of application. Hence, some wise inventors take their whole 6 months for each reply, and make their correspondence sufficiently unsatisfactory that the examiner will answer over and over, in an effort to get the application modified so that it can be approved. In the case of the Linotype patent, no less than 45 years passed between the application and the final granting of a patent, so that Mergenthaler enjoyed a 62 year monopoly on his invention!

Mergenthaler himself died long before his patent was granted, but we may sur-
mise that he did not die unhappily because of this fact.

There are many details concerned with printing which are of interest to editors. Since technically he should select the type sizes and styles to be used, he
should have a type samplebook at hand, and be familiar with the methods u s e d
by printers in specifying sizes. Here is a brief summary of the essential information.

Many systems for measuring type size were devised in the centuries s i n c e
Gutenberg, and the situation was confused by the use of n a m e s f o r different
sizes, such as agate, pica, nonpareil, etc. which represented different sizes in
different countries, and even with different type manufacturers. In 1887, American Type Founders agreed on a unit of measure called the point, which equals
0.01384 inches, or very nearly 1/72 inch. The size of type in <u>points</u> i s t h e
height of the slug required, from the lowest point o f a letter l i k e y, g, or p
(which has a descender) to the highest point of a letter like h, k, o r d (w h i c h
has an ascender). The slug for 10-point type would be 0.1384 inches high, which
is 10 points high.

A horizontal distance equal to the height of type i s called t h e "set em" o r
"em" of the type. For 10-point type, one em is 0.1384 inches long, while f o r
12-point type an em is 0.166 inch, or almost exactly 1/6th inch. This particular sized em (a 12-point m) is called a <u>pica,</u> and is a universal unit u s e d to
measure the length of a line, not only for 12-point type but for any other size as
well.

The following relations exist in these measures:

72 points = 1 inch (almost)

12 points = 1 pica (exactly)

6 picas = 1 inch (almost)

72 picas = 1 foot (almost)

The columns of newspapers are usually about 12 picas wide. If half pica column rules are used between columns, then an 8-column page will be $99\frac{1}{2}$ picas,
or 16.525 inches, wide.

Another unit of length is the "agate line," which is 1/14th inch. It i s u s e d
for the small agate type ($5\frac{1}{2}$ point type) for classified ads.

In the days when type was handset, a typesetter would fill a "stick" or t y p e
holder with about a 2 inch depth (top to bottom of image, as on a page) of type.
The <u>large</u> holders for type used for printing proofs are called <u>galleys,</u> and they
contain 20 inches of type (top to bottom, as on a page).

When the raised portion of type (top to bottom of letter) covers the full width above a letter like k (in the next lower line) might cause the two letters to contact, e. g. , $\underset{k}{y}$. To give more space between lines, the slug can be made larger, e. g. , a 12-point body can be used for 10-point type. Alternately, narrow strips of metal can be inserted as spacers between lines to separate them. These spacers (which do not print) are called leads (because they are made of lead), and the typography is said to be leaded; e. g. , 8-point type may be leaded to lines having 12-point depth (from top to bottom of the metal line as it makes up part of the page).

In giving instructions to a printer, an editor might specify 24 pica lines of 10-point Times Roman type, leaded to 14 points. Or he may say "Times Roman, 10 on 14 by 24," which is usually written, "T. R. $\frac{10}{14}$ x 24 ."

An important problem for a production editor is to estimate the number of printed pages corresponding to a typed manuscript.

The number of characters per line or per page depends on the type size, the length of the printed line in picas and, of course, which letters are used. The latter parameter can be estimated from a table of letter frequency in the English language. Using such a table a tabulation can be made for each size and style of type face. Such tabulations are then printed for common reference by production editors and others in determining how many lines or pages of set type will result from a given piece of copy.

This task of determining the space to be required by the copy when it is set begins with determining the number of characters in the original manuscript. The number of characters in manuscript copy typed on a conventional typewriter can be measured easily, since typewriters have either 10 letters per inch (pica size) or 12 letters per inch (elite size).

An editor can estimate the number of characters on a typed page by multiplying the number of lines by the average length of a line in terms of tenths or twelfths of an inch. For example, if the lines average 6 inches long, then there are 60 pica characters per line or 72 elite characters. If there are 24 full lines on a page, there will be 1440 pica characters on the page, or 1728 elite characters. If this material is to be printed with a 20 pica line in 12-point type , then there will be 48 characters per line, and the 1440 pica type characters will take up 1440/48 = 30 lines of printing. If the 12-point type is printed solidly without added spaces or leading between lines, each line will occupy 12/72 = 0.166 inch, and the 30 lines will be 5 inches deep from top to bottom. The body of solid type is thus 20/6 = 3.33 inches wide by 5 inches deep.

Printing Layout

One of the great advantages of printed copy over a typed manuscript lies in the variety of type style and size available, making possible copy of enhanced readability and attractive design. A typewritten manuscript is an austere and bare desert in comparison to a well planned printed page. The overall effect of a given piece of prose is enormously improved when the material is printed, with justified margins, spacing between paragraphs, and with headings in a different type style. The use of italics and bold-face type provides emphasis, and the justified righthand margin on a printed page produces an impression of professional quality. The production editor has an opportunity, in his coordinated work with the printer, to produce publications of quality and attractiveness. Experience has shown that a proposal which is printed with good style cannot help but create a better impression of a company than a simple mimeographed document can.

Printed publications can be classified into three kinds:

1. Material which people pay to read, such as magazines and books.

2. Material which they must read of necessity, such as income tax forms and timetables.

3. Material which people must be lured into reading, such as advertising.

The greatest efforts in layout to create an attractive overall effect are made in advertisements, because of the intense competition between products. Somewhat less intense, but still serious, is the effort made to make books and magazines easy and pleasant to read. At the bottom of the heap, from the standpoint of layout, we find such items as government forms. These are sometimes dreary wastes of fine print, difficult to understand, and wearying to the most dedicated reader. The writers of this material may make no effort to plan it attractively because they know they have a captive audience -- readers who must struggle through it whether they like it or not.

Well planned ads and books provide the prospective editorial layout man with good examples of typography and arrangement, while the endless numbered forms printed by the bureaucracy provide an inexhaustible supply of horrible examples to avoid.

The technical production editor does not need to worry about the niceties of advertising layout, such as the selection of several sizes and styles of type to lead the eye of a reader unerringly to the product, or the addition of fancy borders and ornaments (called dingbats in the trade). His concern is with the selection of a type size and style for the text, which is easy to read and which will create an impression of stability and dignity.

The first question to answer is, what style and size of type? For the body of the text, the style should be easily read, which means usually a Roman type, neither bold nor lightface, since the bold is reserved for headings or emphasis, and the italics for special words which should be distinguished from the body of the text. For size, 10 point is commonest. Anything smaller than 8-point becomes difficult to read. The use of 12 point, while taking up more space, gives excellent readability -- which will be appreciated by senior executives who may not have as good eyesight as younger men!

The second basic question is, how long should the line be? This depends on the type size, and so the rules of thumb are in terms of size. One rule is as follows: Set a complete alphabet (lower case) from a to z. Measure the length in picas, and then make the line $1\frac{1}{2}$ times this. For textbooks another rule is to use the length of two alphabets. The latter rule is probably best for the text material of a technical report. Another rule which leads to pleasing line length is called doubling the point size. By this rule, the length of the line in picas is made twice the number of points in the type size. Thus for 10-point type, the best line would be 20 picas long ($3\frac{1}{2}$ inches).

Lines that are very short may require excessive spaces between words in order to justify both margins, which makes reading difficult. Very long lines may create confusion for the reader who may have trouble finding where the next line begins.

There are many styles and sizes of type for the printer or editor to choose from. The body of a text is usually printed from a Roman style, like that used in books or newspapers. Letters of the same size and style, but with thicker lines are called boldface or bold. Bold type may be inserted in the text in order to emphasize a word. Italic type has slanting lines, and is used to distinguish a word or phrase, such as a foreign word or expression, or a short quotation, as well as being used for emphasis. Bold type is often used for paragraph headings, sometimes in a larger size than in the text. Boldface type in a different style is often used to make paragraph headings stand out, as is done in most Grantham lessons.

In the body or text, easy readability is the first requirement. Letters that are too small or large, too light or bold, or in an unusual style (as script or black letter) create confusion and should not be used. All-capitals should not be used because words formed from capitals are more difficult to read than lower-case words. The reason lies in the fact that lower-case letters have risers or ascenders as in d, b, and 1, and descenders as in y or j, which give a more distinctive appearance to a word. The relative readability of upper and lower case has received a careful study by state highway departments in designing signs. As a result, all upper case is used for a small vocabulary of standard words, such as STOP, CURVE, DIP, and SPEED LIMIT 25, while words which a driver may see for the first time, such as the name of a town, are painted in lower case for increased readability.

There are many excellent texts on printing layout which give detailed instructions and information for display layout, such as tables of characters per pica for various sizes and styles of type. The editor who studies books of this kind, even when they go beyond his immediate concern on the job, will be able to do a more professional job than the man who has confined his studies to the bare minimum of information needed to prepare a bare technical report.

Summary

In preparing a technical publication, there are four important functions, each

represented by a man having definite areas of talent and responsibility. An engineer usually originates the information contained in a report. A writer puts
the information into an efficiently readable form. The editor sees to it that it
conforms to rules of usage and format, and generally guides the final printed
appearance. Lastly, the printer attends to details of typesetting and the actual
printing by whatever process is chosen.

The editors function is dual, sometimes requiring two separate persons. The
copy editor is concerned with spelling, grammar, good English usage, and conformity with style and format requirements, such as those spelled out in government specifications.

An editor should not be called upon to rewrite major parts of a report, but
when necessary his correction of usage and grammar may require a certain
amount of judicious pruning or, more rarely, addition.

The production editor stands between the writing department, represented by
the writer and copy editor, and the printers who do the actual work of putting
the publication into its finished form. This editor must straddle two technical
worlds. A man with experience in both writing and printing makes the best production editor, with the oldtime tramp printer or country editor who did everything providing the ultimate ideal. A writer or copy editor, however, can soon
learn the printing technology and trade jargon well enough to select type, correct proofs, and do the other required work of coordination.

The qualities needed in an editor are attention to detail, and a kind of intellectual doggedness in being sure that everything is right. Your writer once knew
an outstanding copy editor who was foreign born, and spoke English with an accent, yet had a command of usage that was both scholarly and practical. She
brought to her job the painstaking care that a native English speaking person
would find difficult to accomplish, because she could remember all the rules
that she had to learn, whereas the native speaks intuitively.

Like all middlemen, the editor must be a diplomat. In correcting the brain-
child of a writer, he must exercise tact. He must serve as a kind of soft bump-
er between prima donnas, in the camp of either writers or printers. His job is
to make things go well, rapidly, and smoothly, and his ultimate triumph is a
finished report, proposal, or other publication of which all may be proud.

TEST TC-9

1. Flexible aluminum plates are used in modern offset lithography. . . . . _____

2. An elite typewriter writes 10 characters per inch. . . . . _____

3. A copy editor has less contact with printers than a production editor does. . . _____

4. Writing specifications published by the Department of Defense are suggestive only, and are not binding on writers or editors as to requirements. . . . . . . _____

5. The original writer of a technical manual never does proof reading · · · · · · _____

6. Lithography from stone slabs is the oldest known printing process. . . . . . . _____

7. The Xerox photocopy process competes economically with offset printing for runs of about 25 copies. . . . . _____

8. In offset printing, the copy on the printing plate reads backwards right to left. _____

9. An error in a linotype slug requires recasting the entire line. . . . . _____

10. An editor is intimately concerned with the final layout of a technical report. . _____

MULTIPLE-CHOICE QUESTIONS

11. A pica typewriter is used to type 1400 characters in 22 lines. The average length of a line is
   1. 6.36 inches   3. 7 inches   5. 6.7 inches
   2. 5.3 inches   4. 10.6 inches   _____

12. The number of points (printing unit of measure) in one foot is
   1. 72   2. 864   3. 720   4. 144   5. 12   _____

13. Serifs on letters were invented because
   1. They make the letters set more evenly, by equalizing their height
   2. They improved the appearance of hand-lettered manuscripts
   3. They were used in ancient Norse runes
   4. They make type alignment easier in handset galleys
   5. They eliminate uninteresting and unsightly type frills   _____

14. One of the primary advantage of the linotype process is that
   1. It reduces the type-metal weight
   2. It permits use of a wider variety of type faces than in hand setting
   3. It permits instantaneous correction of typographical errors by just changing one letter in a line
   4. No re-distribution of type to the type cases is required after printing is completed
   5. It can be used for larger type sizes than in hand setting   _____

15. For the body of a technical report, the type style selected for use is most often
   1. Black letter   4. Sans serif
   2. Script   5. Roman
   3. Italics   _____

16. When a proofreader underlines a word on his galley, it means that
    1. The word should be printed in small capitals
    2. The word should be deleted
    3. The word should be printed in italics
    4. The word is misspelled
    5. The word should be printed in bold type

    _____

17. The word "stet" in proofreading means
    1. Let it stand   3. Transpose   5. Straighten the lines
    2. Omit word   4. Wrong type font used

    _____

18. The photo-diffusion process is
    1. Another name for Xerox
    2. Sometimes used in making photo-offset plates
    3. Similar to the ditto process
    4. Dependent upon the application of heat
    5. A process first used in the 17th century

    _____

19. The year 1440 is noted because at this time
    1. Lithography was invented
    2. Woodcuts were first used for illustrations
    3. A book was printed from movable type
    4. Gutenberg was born
    5. The first example of four color printing was made

    _____

20. The production editor for a publication associated with a U. S. Department of Defense contract must
    1. Be good at grammar and spelling
    2. Have a understanding of literary style and format
    3. Be thoroughly familiar with government writing specs
    4. All of the above
    5. None of the above

    _____